T0137034

Studies in Computational Intelligence

Volume 929

Series Editor

Janusz Kacprzyk, Polish Academy of Sciences, Warsaw, Poland

The series "Studies in Computational Intelligence" (SCI) publishes new developments and advances in the various areas of computational intelligence—quickly and with a high quality. The intent is to cover the theory, applications, and design methods of computational intelligence, as embedded in the fields of engineering, computer science, physics and life sciences, as well as the methodologies behind them. The series contains monographs, lecture notes and edited volumes in computational intelligence spanning the areas of neural networks, connectionist systems, genetic algorithms, evolutionary computation, artificial intelligence, cellular automata, self-organizing systems, soft computing, fuzzy systems, and hybrid intelligent systems. Of particular value to both the contributors and the readership are the short publication timeframe and the world-wide distribution, which enable both wide and rapid dissemination of research output.

Indexed by SCOPUS, DBLP, WTI Frankfurt eG, zbMATH, SCImago.

All books published in the series are submitted for consideration in Web of Science.

More information about this series at http://www.springer.com/series/7092

Jongbae Kim · Roger Lee

Editors

Data Science and Digital Transformation in the Fourth Industrial Revolution

 Springer

Editors
Jongbae Kim
Startup Support Foundation
Soongsil University
Seoul, Korea (Republic of)

Roger Lee
ACIS International
Mount Pleasant, MI, USA

ISSN 1860-949X ISSN 1860-9503 (electronic)
Studies in Computational Intelligence
ISBN 978-3-030-64771-1 ISBN 978-3-030-64769-8 (eBook)
https://doi.org/10.1007/978-3-030-64769-8

This Springer imprint is published by the registered company Springer Nature Switzerland AG
The registered company address is: Gewerbestrasse 11, 6330 Cham, Switzerland

Preface

The purpose of International Semi-Virtual Workshop on Data Science and Digital Transformation in the Fourth Industrial Revolution (DSDT 2020) held on October 17, 2020, Soongsil University, Seoul, Korea, is aimed at bringing together researchers and scientists, businessmen and entrepreneurs, teachers and students to discuss the numerous fields of computer science and to share ideas and information in a meaningful way. This workshop on data science and digital transformation in the Fourth Industrial Revolution will discuss the wide range of issues with significant implications, from data warehousing, to data mining, online analytical processing and reporting, data quality assessment, data-driven business models, data privacy and security issues, etc.

This publication captures 17 of the workshop's most outstanding papers. The selected papers have not been published in the workshop proceedings or elsewhere, but only in this book. We await the important contributions that we know these authors will bring to the field.

In Chapter "A Study on the Intention to Use Korean Telemedicine Services: Focusing on the UTAUT2 Model," Harim Byun and Jongwoo Park investigate the factors influencing the intention to use Korean telemedicine services from the perspective of potential overseas customers. This study has tested the influence relationship by reconstructing the UTAUT2 model to be suitable for the studies on the acceptance of telemedicine services.

In Chapter "A Structural Relationship Between Environmental Uncertainty, Dynamic Capability, and Business Performance in a Smart Supply Chain Environment," Yongmuk Kim and Jongwoo Park tested their hypothesis using structural equation model analysis to verify the structural relationship between dynamic capability and business performance in the smart supply chain environment of Korean small and medium manufacturers. They confirmed through this that dynamic capability and environmental uncertainty in a smart supply chain environment have a partially significant effect on business performance and that manufacturers must strive to implement direct, systematic policies and improve dynamic capability to respond to an uncertain business environment.

In Chapter "Analysis of the Relative Importance of Coaching Service Quality," Jinyoung Yang, Hyoboon Wang and Donghyuk Jo analyze the relative importance between determinants and factors of coaching service quality. Result of this study may be used as basic data in establishing an academic knowledge system of coaching and expanding the base for coaching.

In Chapter "Determinants of Technology-Based Self-Service Acceptance," Seulki Lee and Donghyuk Jo explore the factors that effects the acceptance intention of technology-based self-service of food and beverage store customers. The results of this study have academic implications for investigating the antecedent variables to the acceptance intention of technology-based self-service in uncertain environments. In practice, it will contribute to decision-making and business strategy formulation for the post-corona era.

In Chapter "Pre-verification of Data in Electronic Trade Blockchain Platform," Saeyong Oh, Sanghyun Cho, Sunghwa Han and Gwangyong Gim propose an architecture that can verify the integrity of the blockchain-based electronic document platform by presenting a pre-validation method for electronic documents.

In Chapter "Study on Security and Privacy of E-Government Service," Sanghyun Cho, Saeyong Oh, Hogun Rou and Gwangyong Gim analyzed prior studies and cases of e-government in order to derive the factors influencing the continuous use of e-government services provided by central government departments, local governments and public institutions, and to establish the causal relationship between each factor.

In Chapter "Applying Hofstede's Culture Theory in the Comparison between Vietnam and Korean E-government Adoption," Hung-Trong Van, Simon Gim, Euntaek Lim and Thi-Thanh-Thao Vo find the factors affecting the intention to use e-government by comparing Korea and Vietnam citizens' behavior based on TAM and IS success model.

In Chapter "Study on Business Strategy Quantification using Topic Modeling and Word Embedding: Focusing on 'Virtual Reality' and 'Augmented Reality'," Siyoung Lee, Sungwoong Seo, Hyunjae Yoo and Gwangyong Gim analyzed articles related to "virtual reality" and "augmented reality," the main elements of the Fourth Industrial Revolution, and reviewed the views that the media has by using text mining techniques. This study showed that topics derived from topic modeling can be expressed in strategic space both nominally and visually.

In Chapter "3D Printing Signboard Production Using 3D Modeling Design," Jungkyu Moon and Deawoo Park proposed 3D printing signboard production using 3D modeling design. The signboard studied in this paper has the advantages of long life and low failure rate, and the disadvantage of high cost is to be solved by using fourth industrial technology.

In Chapter "AI-Based 3D Food Printing Using Standard Composite Materials," Hyunju Yoo and Daewoo Park proposed AI-based 3D food printing using standard composite materials. This study overcome the limitation of 3D food printing, which cannot have standardized design values for printability due to the variety of food raw materials and materials added to improve printability.

In Chapter "Telemedicine AI App for Prediction of Pets Joint Diseases," Suyeon Han and Deawoo Park proposed a smartphone app based on 5G communication that can diagnose the presence and possibility of bone joint disease in companion animals.

In Chapter "Design of Artificial Intelligence for Smart Risk Pre-review System at the KC EMC," Youngjoo Oh and Deawoo Park proposed an AI deep learning system that provides a complementary measure and provide it in the form of a cloud computing platform. And they proposed a semi-automatic AI connection system that can perform KC conformity assessment for electric and electronic products to be sold in Korea.

In Chapter "AI Analysis of Illegal Parking Data at Seocho City," Donghyun Lim and Deawoo Park proposed an AI machine learning system that links the vehicle's number recognition algorithm for illegal parking. And they suggest an advanced system that analyzes the status of illegal parking and stopping judgment in Seocho City Office, where big data and AI are connected using spatial information and AI.

In Chapter "Regularized Categorical Embedding for Effective Demand Forecasting of Bike Sharing System," Sangho Ahn, Hansol Ko and Juyoung Kang propose a regularized categorical embedding methodology that can learn not only the station-centric model but also the overall trend to predict the rental demand for shared bicycles. The proposed methodology is expected to be an effective prediction for station-centric models with unbalanced performance distribution and the prediction of rental demand for shared bicycles.

In Chapter "Development of a Model for Predicting the Demand for Bilingual Teachers in Elementary Schools to Support Multicultural Families—Based on NEIS Data," Jinmyung Choi and Dooyeon Kim developed a test model using big data to analyze the data of school sites and data held by administrative agencies and applied the actual data to this model to analyze the policy's appropriateness and effectiveness. This study's results can help establish and implement policies in various educational fields through big data analysis.

In Chapter "Research on Implementation of User Authentication Based on Gesture Recognition of Human," Jungseon Oh, Joongyoung Choi, Kwansik Moon and Kyoungho Lee propose a method to implement such biometric authentication based on gestures (dynamic characteristics) and body characteristics (static characteristics) and suggest ways to use them.

In Chapter "AI TTS Smartphone App for Communication of Speech Impaired People," Hanyoung Lee and Deawoo Park proposed AI TTS smartphone app for communication of speech impaired people. This study will be used as a means of transmitting information to the hearing impaired in the untact era.

It is our sincere hope that this volume provides stimulation and inspiration, and that it will be used as a foundation for works to come.

Seoul, Korea (Republic of)   Jongbae Kim
August 2020

Contents

Contributors

Sangho Ahn Department of e-Business, Ajou University, Suwon, Republic of Korea

Harim Byun Soongsil University, Seoul, South Korea

Sanghyun Cho Department of IT Policy and Management, Soongsil University, Seoul, South Korea

Jinmyung Choi Department of Convergence IT, Konyang University, Chungchungnam-Do, Republic of Korea

Joongyoung Choi Defense Information System Management Group, KIDA, Seoul, Republic of Korea

Gwangyong Gim Department of IT Policy and Management, Soongsil University, Seoul, South Korea;
Department of Business Administration, Soongsil University, Seoul, Korea

Simon Gim SNS Marketing Research Institute, Soongsil University, Seoul, Korea

Sunghwa Han Department of IT Policy and Management, Soongsil University, Seoul, South Korea

Suyeon Han Department of Convergence Engneering, Hoseo Graduate School of Venture, Seoul, Korea

Donghyuk Jo Department of Business Administration, Soongsil University, Seoul, South Korea

Juyoung Kang Department of e-Business, Ajou University, Suwon, Republic of Korea

Dooyeon Kim Department of Convergence IT, Konyang University, Chungchungnam-Do, Republic of Korea

Yongmuk Kim Soongsil University, Seoul, South Korea

Hansol Ko Department of e-Business, Ajou University, Suwon, Republic of Korea

Hanyoung Lee Department of Convergence Engineering, Hoseo Graduate School of Venture, Seoul, South Korea

Kyoungho Lee Graduate School of Information Security, Korea University, Seoul, Republic of Korea

Seulki Lee Department of Business Administration, Sangmyung University, Seoul, Korea

Siyoung Lee Department of IT Policy and Management, Graduate School, Soongsil University, Seoul, Korea

Donghyun Lim Department of Convergence Engineering, Hoseo Graduate School of Venture, Seoul, Korea

Euntaek Lim Graduate School of Business Administration, Soongsil University, Seoul, Korea

Jungkyu Moon Department of Convergence Engineering, Hoseo Graduate School of Venture, Seoul, Korea

Kwansik Moon National Assembly, Seoul, Republic of Korea

Jungseon Oh KEPCO, ICT Planning, Seoul, Republic of Korea

Saeyong Oh Department of IT Policy and Management, Soongsil University, Seoul, South Korea

Youngjoo Oh Department of Convergence Engineering, Hoseo Graduate School of Venture, Seoul, Korea

Daewoo Park Department of Convergence Engineering, Hoseo Graduate School of Venture, Seoul, Korea

Jongwoo Park Department of Business Administration, Soongsil University, Seoul, South Korea

Hogun Rou Department of IT Policy and Management, Soongsil University, Seoul, South Korea

Sungwoong Seo Hanwha Systems, Seoul, Korea

Hung-Trong Van Faculty of Digital Economy & E-Commerce, Vietnam–Korea University of Information and Communication Technology, Danang, Vietnam

Thi-Thanh-Thao Vo Faculty of Digital Economy & E-Commerce, Vietnam–Korea University of Information and Communication Technology, Danang, Vietnam

Hyoboon Wang Soongsil University, Seoul, South Korea

Jinyoung Yang Soongsil University, Seoul, South Korea

Hyunjae Yoo Department of IT Policy and Management, Graduate School, Soongsil University, Seoul, Korea

Hyunju Yoo Department of Convergence Engineering, Hoseo Graduate School of Venture, Seoul, Korea

A Study on the Intention to Use Korean Telemedicine Services: Focusing on the UTAUT2 Model

Harim Byun and Jongwoo Park

Abstract With the outbreak of the recent COVID-19 pandemic, the importance and demand of telemedicine, a method of supply medical services in a non-face-to-face manner, is spreading worldwide. Despite its high medical standards and IT technology, the telemedicine practice between medical staff and patients is not allowed in Korea. Therefore, this study was intended to examine the factors influencing on the intention to use Korean telemedicine services from the perspective of potential consumers in China and Vietnam as a pavement basic work to expand Korean telemedicine services overseas. To this end, the UTAUT2 model was reconstructed to be suitable for the study on the acceptance of telemedicine service; users' perceived perception on Korean telemedicine services (i.e. performance expectancy, effort expectancy, social influence, price value, perceived risk) and personal characteristics (i.e. innovativeness and concern for health) were set as main variables; and their relationship with the use intention was empirically analyzed. As a result of the analysis, it was found that, performance expectancy, social influence, price value, and perceived risk factors, except for effort expectancy, have a significant relationship influencing on the use intention. In addition, it was confirmed that innovativeness and concern for health have a moderating effect on the relationship between performance expectancy and use intention. Finally, specific implications derived based on the analysis results are expected to be used as basic data for seeking global strategies for Korean medical services in conjunction with the overseas expansion of the Korean telemedicine service-related businesses.

Keywords COVID-19 · Telemedicine · UTAUT2 · Korean telemedicine service · Telehealth · Global medical service · Perceived risk · Use intention

H. Byun
Soongsil University, Seoul, South Korea
e-mail: cou80@naver.com

J. Park (✉)
Department of Business Administration, Soongsil University, Seoul, South Korea
e-mail: jongpark7@ssu.ac.kr

© The Author(s), under exclusive license to Springer Nature Switzerland AG 2021
J. Kim and R. Lee (eds.), *Data Science and Digital Transformation in the Fourth Industrial Revolution*, Studies in Computational Intelligence 929,
https://doi.org/10.1007/978-3-030-64769-8_1

1 Introduction

The recent outbreak of COVID-19 pandemic is accelerating a "zero contact society" worldwide. Under the main word of 'isolation and closure', we are experiencing changes in all lifestyles and industries (e.g. consumption, distribution, education, etc.). In particular, as the necessity and market demand for the virtual telemedicine in which medical staff and patients do not face each other have been confirmed, the virtual telemedicine is rapid spreading worldwide, especially in the advanced [1].

Global Market Insights reported that the scale of the global digital healthcare market is expected to grow from KRW130 trillion (USD16.4 Billion) in 2019 to KRW600 trillion (USD5.404 Billion) in 2025, by an annual average of 29.6%; and the scale of telemedicine market is also expected to reach $32.71 Billion by 2027 [2].

With its advantages, such as improving access to medical services and reducing medical costs in addition to the prevention of infection on account of its non-contact method, the telemedicine has emerged as a solution to the rapid increase in medical expenses caused by the aging of the population, and the number of corporations and governments in many countries promoting the telemedicine is increasing rapidly. In particular, in countries with high-level medical and IT technologies (e.g. the United States, Europe, and Japan), the telemedicine service is being promoted in response to growing demand for efficient health care, and various business models are being developed.

Korea has demonstrated its remarkable response capabilities based on rapid diagnosis, thorough quarantine and high-level medical standards during this COVID-19, gaining credibility worldwide. As a result, K-bio and K-medicine of Korea in conjunction with Korean Wave have gained international credibility. In addition, Korea, with high-level information and communication technology enough to commercialize 5G for the first time in the world, has the best conditions to provide telemedicine services beyond doubt. However, Korea is the only country in the world that telemedicine practice between medical staff and patients is not allowed to date for the reasons of the issues related to proving the effectiveness and safety of telemedicine for many years [3]. In addition, on account of various legal regulations, the development of related technologies, the scale of related corporation and investments and the growth rate of corporation concerned are very low compared to other countries. Now, the telemedicine service attracts intense attention as "must-have" rather than "good-if-we-have" service before and after the outbreak of the global epidemic. In order not fall behind such an inevitable global change, it is necessary to first develop it as a global medical service business by turning our eyes to overseas markets where telemedicine service can be more freely implemented, not to the domestic market. In particular, telemedicine service can be intensely used as a strategic means to attract overseas patients into Korea and to increase the possibility of the overseas expansion of Korean medical institutions and medical consultations for potential medical tourists are intensely being conducted remotely in preparation for the Post-COVID.

Studies on the acceptance and use of new information technologies are said to be an advanced study field, but studies on the acceptance of technologies in the health care have relatively seldom been conducted [4]. In particular, for the telemedicine, studies have mainly focused on the possibility of technical implementation to date, and studies on the inducement of practical use of technologies have been conducted mostly on the Koreans for whom it is difficult to apply the telemedicine in Korea, so they are considered to have limitations in light of the effectiveness of studies [5]. Therefore, this study has focused on China and Vietnam as main overseas market for telemedicine service at the present time after thoroughly reviewing the rankings of medical tourists visiting Korea by country, the status of the overseas expansion of Korean medical institutions and the constraints on overseas telemedicine services (i.e. accessibility, time difference between countries and the extent of telemedicine activa-tion within the country), conducting an empirical on the potential customers in those countries. In addition, this study intends to discover the acceptance factors related to Korean telemedicine service and confirm the causal relationship with the use inten-tion by using UTAUT (Unified Theory of Acceptance and Use of Technology). The ultimate purpose of this study is to predict the possibility of the success of Korean telemedicine service business in overseas markets; derive matters to supplement; and ultimately provided basic data needed for seeking strategies to globalize Korean medical services, including medical tourism.

2 Theoretical Background

2.1 The Concept and Definition of Telemedicine Service

There is no unified definition of the concept of telemedicine. In general, various terminologies (i.e. Telemedicine, Telehealth, Telehealthcare, e-health and U-health) are currently used interchangeably.

When classifying the concepts of the 'telemedicine' and the 'telehealth', which are being most commonly used interchangeably, the telemedicine is a concept that includes telehealth, and the telehealth occupies part of the telemedicine. The telemedicine can be understood as the remote replacement of the treatment of patients by doctors in an in-hospital clinic through various communication technologies (e.g. video treatment, telephone treatment, remote secondary medical opinion and remote prescription, etc.). On the other hand, the telemedicine is a comprehensive concept that includes both remote patient monitoring and remote surgery, which are services that remotely monitor patient health and disease in addition to treatment-related services [3].

When looking at a few of conceptual definitions of telemedicine, Krupinski et al. [6] defined the telemedicine as the "exchange of patient-related health information between geographically distant medical service providers or between providers and consumers (patients) through telecommunication technology and computers for the purpose of evaluation, diagnosis, treatment and education", and the World Medical Association (WMA) defined telemedicine as "a medical practice that determines and recommends intervention, diagnosis and treatment for diseases based on clinical data, records and other information transmitted from a distance through a telecommunication system". Also the World Health Organization (WHO) defined the telemedicine as the "act of all medical experts to prevent, diagnose, treat diseases or injuries; continuously educate medical service providers; and provide useful information and medical services for communities and local residents from distant places using information and communications technologies" [7].

When looking at these definitions, the telemedicine is viewed differently, focusing on the central functions performed (e.g. viewing the telemedicine as an alternative means of face-to-face treatment and viewing it as exchange of technology and medical information), but what is common between various definitions is viewing the telemedicine as the combination of medical service and information and communication technology. This study views the telemedicine from a comprehensive perspective and defines the telemedicine as all medical-related activities performed in a noncontact manner using information and communication technology between various telemedicine service providers and telemedicine service recipients.

2.2 The Current Status of Overseas Telemedicine Service Businesses at Home and Abroad

Thailand, Singapore, and India leading the global medical tourism market have introduced the telemedicine from the beginning of 2000 as a countermeasure in response to the gradual decline in the profits from medical tourists [8]. Table 1 briefly shows the

Table 1 Overseas telemedicine service status

Country (hospital)	Contents
Thailand (Bumrungrad Hospital)	Through "Global Care Solution (GCS)" project (i.e. an online pre-treatment) promoted in collaboration with MS from Dec 2007, it has strengthened its pre-treatment function, shortening waiting time
Singapore (Raffles Hospital)	Through the operation of an online communication channel called "ASK RMG" since 2008, medical staff responds in detail to inquiries about health care from overseas customers in real time. This promotes hospitals naturally, resulting in an increase in the number of recruited customers
India (Apollo Hospital)	Founding the "Remote Medical Network", it has established telemedicine service cooperation system in 9 countries (e.g. Southwest/Central Asia, including Sri Lanka, Pakistan, Nepal, Bangladesh, Myanmar, and Africa)

Source Taegyu et al. [8]

current status of the telemedicine services at the representative hospitals attracting overseas patients in Thailand, Singapore and India.

In addition, Cuba is providing telemedicine services to countries in Central America and the Caribbean, and Mexico has been conducting international telemedicine services through cooperation with specific medical institutions in the United States by intensively investing in the telemedicine infrastructure since 2014.

In the early 2000s, Korea's overseas telemedicine service business began as part of Official Development Assistance (ODA) project for countries with underdeveloped health care environment, and as a result, the number of successful cases attracting and treating patients in Korea has gradually increased [9].

As a result of the successful operation of the U-Health system, a remote video treatment system installed in Vladivostok, Russia in 2011, Gangnam Severance implements the personalized medical services for overseas patients, promoting the recruitment of patients.

Seoul National University Hospital has been actively implementing remote reading and remote consultations that receive medical opinions and consultation from Seoul National University Hospital through a computational monitoring system (PPCC: Pre-post Care Center) installed in the Sheikh Khalifa Hospital in the United Arab Emirates, whose operation was commissioned to Seoul National University Hospital from 2015. As above, Korean telemedicine service is generally promoted by the connection between overseas governments and specific large domestic hospitals under the policy intended to expand medical care for overseas patients. Large domestic hospitals provide remote medical services similar to telemedicine services by opening telemedicine centers through medical tourism centers and local hospitals located overseas. Currently, domestic small and medium-sized hospitals (e.g. plastic surgery, dermatology and oriental medicine clinics) mainly focus on remote consultation using phone, e-mail, and social media for the purpose of attracting overseas patients into Korea [10].

2.3 United Theory of the Acceptance and Use of Technology

The first theory intended to explain the phenomena of accepting new technologies is the Technology Acceptance Model (TAM) suggested by Davis [11]. This model was developed to apply the relationship between beliefs, attitudes, intentions and actions to the acceptance of technology based on theory of rational behavior. The TAM has been verified as a representative model with high explanatory power in numerous technology acceptance studies since it was suggested. On the other hand, controversy over the limitations of the proposed model in the organizational context and the inability to sufficiently reflect the influence of various exogenous factors has been raised steadily. As a supplement to this, Venkatesh et al. [12] suggested a new technology acceptance model (UTAUT: Unified Theory of Acceptance and Use of Technology) from an integrated perspective, but this model also focuses on the factors influencing organization members' acceptance of information technology.

Since then, Venkatesh et al. [13] also presented the UTAUT2 (Extended Unified Theory of Acceptance and Use of Technology) model that could better explain the process of accepting and using technology by general consumers. In UTAUT2 model, 3 factors (i.e. hedonic motivation, price value and habit) were added to the 4 core factors of the existing UTAUT model (i.e. performance expectancy, effort expectancy, social influence and facilitating conditions) and these factors have been confirmed to be critical variables in the results of recent researches using these factors [14–17].

2.4 Perceived Risks

The perceived risk initially suggested by Bauer [18] is the risk perceived subjectively in the situation where consumers have multiple choices, referring to the uncertainty and negative consequences that consumers feel when they cannot predict the outcome of their actions. In other words, psychological discomfort and anxiety occurring as a result of consumers' perception of risk negatively influence the evaluation of product or service, which may, in turn, directly influence the acceptance and purchase of products and services [19]. Prior studies on the acceptance of technology applying perceived risk suggested that the perceived risk reduces perceived usefulness and intention to use and adopt, supporting the above argument [20–22]. The influence of risk perception is more prominent in the situation of the purchase of services than in tangible products and medical services are particularly high-risk services that make it difficult to predict the quality of services until they are experienced [23]. The telemedicine services using various information and communication technologies as a medium are basically exposed to an open environment of so-called online, which may make security main risk factors (e.g. hacking, errors, and personal (biometric) information leakage). In addition, on account of the non-contact interactions between medical staff and patients, the accuracy of communication, diagnosis, prescription and resulting problem-solving issues may act as risk factors.

3 Research Model and Hypotheses

This study intends to investigate the factors influencing the intention of potential overseas customers to use the telemedicine services to expand telemedicine services, which are spreading as a new method of supplying medical services, into overseas markets. Therefore, in this study, the UTAUT2 model was selected as a basic theoretical framework and new factors that could influence the use intention were added to design a model for accepting Korean telemedicine services.

In this study, the 5 main factors set to influence the intention to use Korean telemedicine services were 'performance expectancy', 'effort expectancy', 'social influence', 'price value' and 'perceived risk'. But the 'hedonic motivation' and 'habit'

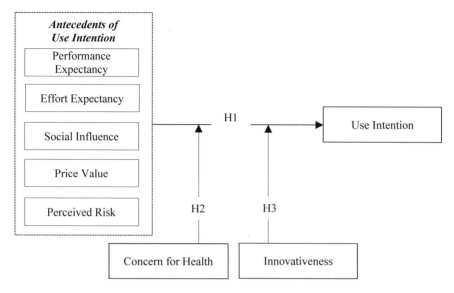

Fig. 1 Research model

factors in the UTAUT2 model were not reflected in this study because it was considered that the level of technological support expected at a personal level would not greatly promote or delay consumers' use of telemedicine on account of the service supply path provided through medical institutions and telemedicine centers equipped with many telemedicine systems, except for medical consultations that consumers mainly use using mobile. In addition, 'hedonic motivation' and 'habit' were eliminated from the influencing factors in consideration of the essential characteristics of medical treatment directly associated with human health and the situational characteristics of the commercialization stage of telemedicine. As there are a number of potential risk factors (e.g. threats to the health rights of patients and leakage of personal (biometric) information due to remote medical services), which was predicted to directly influence the user's intention to use, the perceived risk factors were added to the influencing factors for analysis. Figure 1 shows the research model of this study.

4 Research Method

4.1 Data Collection and Analysis Method

In this study, general consumers in China and Vietnam were selected as potential users of Korean telemedicine services, and a survey was conducted using a random sampling method. The online survey was conducted for two weeks starting from

Table 2 Sample characteristics

Item		Sample size	Ratio (%)	Total
Country	China	151	48.7	310
	Vietnam	159	51.3	
Gender	Male	111	33.4	310
	Female	199	59.9	
Age	20's	143	43.1	310
	30's	126	38.0	
	40's	36	10.8	
	More than 50	5	1.5	
Final education	High school	32	9.6	310
	College	58	17.5	
	University	149	44.9	
	Graduate school	71	21.4	
Job	Office work	79	23.8	310
	Production occupation	17	5.1	
	Professional occupation	19	5.7	
	Student	105	31.6	
	Official	55	16.6	
	Self-employed	35	10.5	

Jul 2, 2020, and finally 310 copies of data were used for analysis. In addition, 7-point Likert scale was used to measure the responses to questions and SPSS 23.0 was used for statistical analysis. The demographic characteristics of the sample group are summarized in Table 2.

4.2 Operational Definition of Variables

This study has selected 'performance expectancy', 'effort expectancy', 'social influence', 'price value' and 'perceived risk' as the main factors influencing the intention of overseas consumers to use Korean telemedicine services through literature review and listening to expert opinions; analyzed the relationship between the factors; and additionally analyzed the moderating effect of personal characteristics (i.e. innovativeness and concern for health) between these factors and the use intention. The operational definitions of the main variables selected in this study are listed in Table 3.

Table 3 Operational definition of variables

Variable	Operational definition	Source
Performance expectancy	The extent to which the use of Korean telemedicine service is believed to be able to help achieve the expected purpose of use	[12, 24]
Effort expectancy	The extent to which Korean telemedicine services are believed to be easy to use	[12, 25]
Social influence	The degree of belief that the reference group will consider Korean telemedicine services positive and recommend to their acquaintances the use of Korean telemedicine services	[12, 26]
Price value	The extent of satisfaction from using Korean telemedicine services compared to the cost to pay for the use	[14, 27]
Perceived risk	The possibility of perceived loss from using Korean telemedicine services	[28, 29]
Innovativeness	Individual's disposition or willingness to quickly experience new technologies and services	[20, 30]
Concern for health	The importance of health and the extent of concern for health that individuals feel	[31, 32]
Use intention	The intention or plan to use Korean telemedicine services now or in the future	[11, 12]

5 Research Results

In this study, multiple regression analysis was conducted to test the hypotheses, and the analysis results are listed in Tables 4 and 5. As a result of the analysis, it was found that the performance expectancy, social influence, price value, and perceived risk had a significant influence on the intention to use Korean telemedicine services. In addition, as a result of analyzing the moderating effects of innovativeness and concern for health, it was found that both innovativeness and concern for health had

Table 4 Results of hypotheses tests

Factor	Use intention			
	B	β	t	p
Performance expectancy	0.055	0.176	3.764**	0.000
Effort expectancy	0.054	0.031	0.571	0.569
Social influence	0.057	0.305	5.273**	0.000
Price value	0.057	0.344	6.094**	0.000
Perceived risk	−0.039	−0.110	−2.959**	0.003
R^2	0.650			
F	112.418**			

Note $*p < 0.05, **p < 0.01$

Table 5 Results of moderating effect analysis

Path	Moderator	β	t	p
PE → UI	Innovativeness	0.657	2.968**	0.003
EE → UI		0.073	0.323	0.747
SI → UI		0.101	0.444	0.657
PV → UI		0.001	0.003	0.998
PR → UI		−0.332	−1.274	0.204
PE → UI	Concern for health	0.572	2.466*	0.014
EE → UI		0.054	0.250	0.803
SI → UI		−0.046	−0.213	0.831
PV → UI		−0.099	−0.424	0.672
PR → UI		−0.521	−1.873	0.062

Note $*p < 0.05$, $**p < 0.01$

a moderating effect on the relationship between performance expectancy and use intention (t = 2.968, $p = 0.003$/t = 2.466, $p = 014$).

6 Conclusions

With the outbreak of COVID-19, the telemedicine is attracting more attention than ever as it emerges as a safe and efficient method of supplying medical services. As a basic study for the overseas expansion and promotion of telemedicine services that are not allowed in Korea, this study was intended to investigate the factors influencing the intention to use Korean telemedicine services from the perspective of potential overseas customers. To this end, this study has tested the influence relationship by reconstructing the UTAUT2 model suggested by Venkatesh et al. [13] to be suitable for the studies on the acceptance of telemedicine services. The analysis results of this study can be summarized as follow.

First, it was found that performance expectancy, social influence, price value, and perceived risk had a significant influence on the intention to use Korean telemedicine services. The significant influence relationship between performance expectancy and intention to use has been proven in various studies on the acceptance of information technology including the telemedicine [29, 33, 34]. Therefore, it deems necessary to actively promote the usefulness of Korean telemedicine services and develop its image based on the advices of medical experts and tests. In addition, management and response strategies to remove or mitigate risk factors must be established within the organization through the process of identifying risk factors perceived by consumers in detail and predicting the consequences of risk according to the source.

Second, it was found that price value and social influence, in particular, greatly influence the intention to use. Since these results indicate that the opinions of neighbors and recommendation to them can be effectively used in the strategies for promoting Korean telemedicine services, it is necessary to promote marketing strategies that encourage sharing experience through Internet media and actively support acquaintance recommendation programs. In addition, it was also found that overseas consumers feel sensitive to the cost they will pay for using Korean telemedicine services. This implies that establishing a rational cost strategy through meticulous comparison of medical expenses in the local market and the telemedicine service products provided in competing countries will become a key factor in expanding the bases of overseas markets.

Third, it was found that the effort expectancy did not significantly influence the intention to use Korean telemedicine services, being consistent with the results of the study conducted by Choi et al. [16]. Given the fact that many of the survey subjects are potential customers who have not actually used telemedicine services and most of the telemedicine services, except for mobile medical consultations, are provided at special institutions (e.g. medical institutions and telemedicine centers), the difference in the extent of individual consumer's effort to use the telemedicine service is considered to be less relevant to the actual use.

Fourth, it was found that concern for health and innovativeness have a moderating effect on the relationship between performance expectancy and intention to use Korean telemedicine services. Accordingly, the need for an opportunity to directly and indirectly experience Korean telemedicine through international fairs related to beauty and health and promotion strategies for Korean telemedicine service products that target users of healthcare-related devices and platforms is emphasized.

For the limitations of this study and future research directions are as follows. First, since a comprehensive survey was conducted without distinguishing the nationality (China, Vietnam) of the survey subjects, this study has limitations in generalizing the study results. Second, this study has limitations in that it did not comprehensively reflect various factors influencing the intention to use Korean telemedicine services, an extended study to supplement these limitations will be required in the future.

References

1. Ahn S, Kim D, Kim B, Choi A (2021) US equity small cap inside 2021. YemoonBook
2. Stratistics Marker Research Consulting (2020) Telehealth-global market outlook (2018–2027)
3. Choi Y (2020) Digital health care: medical future. CloudNineBooks
4. Aggelidis VP, Prodromos DC (2009) Using a modified technology acceptance model in hospitals. Int J Med Inform 78(2):115–126
5. Kim S (2010) Structural relationships among factor to adoption of telehealth service by health conditions. Department of Public Health Graduate School, Inje University
6. Krupinski E, Nypaver R, Poropatich D, Ellis RS, Hasan S (2002) Clinical application in telemedicine/telehealth. Telemedicine J e-Health 8(1):13–34
7. World Health Organization (2010) Telemedicine: opportunities and developments in member states. Report on the second global survey on eHealth. Global observatory for eHealth series, vol 2, Geneva, Switzerland

8. Yu T, Choi Y (2014) Telemedicine centers as a venue for attracting international medical tourists. Korean Public Manage Rev 28(3):133–167
9. Kim S (2017) Case study on problems and solutions of international telemedicine system in Korean general hospitals. Graduate School of Public Administration Seoul National University
10. KukiNews (2020) Hospitals for remote counseling for overseas patients
11. Davis FD (1989) Perceived usefulness, perceived ease of use, and user acceptance of information technology. MIS Q:319–340
12. Venkatesh V et al (2003) User acceptance of information technology: toward a unified view. MIS Q 27(3):425–478
13. Venkatesh V, James YLT, Xin X (2012) Consumer acceptance and use of information technology: extending the unified theory of acceptance and use of technology. MIS Q 36(1):157–178
14. Agarwal NK et al (2007) Factors affecting 3G adoption: an empirical study. PACIS 2007 Proc 3
15. Choi B, Kim H, Chung J (2019) Consumer acceptance of mobile gift certificates-focused on UTAUT2. J Dig Converg 17(9):97–104
16. Choi W, Kim D, Choi S (2017) Understanding factors influencing usage and purchase intention of a VR device: an extension of UTAUT2. J Korea Soc Comput Inf 18(3):173–208
17. Escobar-Rodríguez T, Elena CT (2013) Online drivers of consumer purchase of website airline tickets. J Air Transp Manage 32:58–64
18. Bauer RA (1960) Consumer behavior as risk taking. In: Hancock R (ed) Proceedings of the 43rd America Marketing Association, pp 384–398
19. Featherman MS, Paul AP (2003) Predicting e-services adoption: a perceived risk facets perspective. Int J Hum Comput Stud 59(4):451–474
20. Lu H, Chin-Lung H, Hsiu-Ying H (2005) An empirical study of the effect of perceived risk upon intention to use online applications. Inf Manage Comput Secur 13(2):106–120
21. Crespo ÁH, Del Bosque IR, de los Salmones Sánchez MG (2009) The influence of perceived risk on Internet shopping behavior: a multidimensional perspective. J Risk Rese 12(2):259–277
22. Kim H, Oh S (2019) Effects of gender difference and perceived risk on acceptance intention of mobile easy payment service. J Manage Econ 41(3):145–165
23. Murray KB (1991) A test of services marketing theory: consumer information acquisition activities. J Market 55(1):10–25
24. Kim S, Lee C (2013) Usage intention of u-healthcare service using unified theory of technology adoption and usage. J Korea Contents Assoc 13(12):379–388
25. Rogers EM (2003) Diffusion of innovations, 5th edn. The Free Press, New York
26. Son H, Lee S, Cho M (2014) Influential factors of college students' intention to use wearable device-an application of the UTAUT2 model. Korean J Commun Inf 68:7–33
27. Khan IU, Zahid H, Safeer UK (2017) Understanding online banking adoption in a developing country: UTAUT2 with cultural moderators. J Glob Inf Manage (JGIM) 25(1):43–65
28. Basoglu N, Daim TU, Topacan U (2012) Determining patient preferences for remote monitoring. J Med Syst 36(3):1389–1401
29. Rho M (2013) A study of the expected benefit and perceived risk on telemedicine acceptance for hypertension management. J Int Trade Commerce 9(1):343–361
30. Agarwal R, Jayesh P (1998) A conceptual and operational definition of personal innovativeness in the domain of information technology. Inf Syst Res 9(2):204–215
31. Jayanti RK, Alvin CB (1998) The antecedents of preventive health care behavior: an empirical study. J Acad Mark Sci 26(1):6–15
32. Shin M, Lee Y (2015) A study on the influential factors of purchase intention of wrist wearable device. J Korea Contents Assoc 15(5):498–506
33. Chau PYK, Paul JH (2002) Examining a model of information technology acceptance by individual professionals: an exploratory study. J Manage Inf Syst 18(4):191–229
34. Lanseng EJ, Tor WA (2007) Electronic healthcare: a study of people's readiness and attitude toward performing self-diagnosis. Int J Serv Indust Manage 18(4):394–417

A Structural Relationship Between Environmental Uncertainty, Dynamic Capability, and Business Performance in a Smart Supply Chain Environment

Yongmuk Kim and Jongwoo Park

Abstract E-commerce, based in the 4th industrial revolution to respond to environmental changes such as rapid technical development, increasing global competition, and diverse customer needs, is accelerating while uncertainty in the business environment grows more severe. In order to effectively respond to this uncertain business environment, individual companies and parts of supply chains need flexibility, and this need for dynamic flexibility is growing as uncertainty increases. In this study, we tested our hypothesis using structural equation model analysis to verify the structural relationship between dynamic capability and business performance in the smart supply chain environment of Korean small and medium manufacturers. We confirmed through this that dynamic capability and environmental uncertainty in a smart supply chain environment have a partially significant effect on business performance and that manufacturers must strive to implement direct, systematic policies and improve dynamic capability to respond to an uncertain business environment.

Keywords Smart supply chain · Dynamic capability · Environmental uncertainty · Business performance

1 Introduction

As industrial activity is paralyzed due to recent rapid, dynamic changes in the corporate environment, protectionism, epidemic status, etc., we have been facing crisis not only in our daily lives but worldwide. In particular, untact, i.e. contactless, technology is being promoted throughout society due to COVID-19. E-commerce based on the 4th industrial revolution is accelerating to respond to these changes, and uncertainty in the business environment is growing. Studies proposing a clear definition of the 4th

Y. Kim
Soongsil University, Seoul, South Korea
e-mail: ymkim9471@naver.com

J. Park (✉)
Department of Business Administration, Soongsil University, Seoul, South Korea
e-mail: jongpark7@ssu.ac.kr

© The Author(s), under exclusive license to Springer Nature Switzerland AG 2021   13
J. Kim and R. Lee (eds.), *Data Science and Digital Transformation in the Fourth Industrial Revolution*, Studies in Computational Intelligence 929,
https://doi.org/10.1007/978-3-030-64769-8_2

industrial revolution in e-commerce have been few, but the 4th industrial revolution features a high level of connectivity and superintelligence, meaning that everything is connected by fusing digital devices, humanity, and physical environments to evolve into an intellectualized society [1].

Despite this technical development, businesses' supply chain structure has comprised various participants from raw material producers to component producers, finished item producers, wholesalers and retailers, and transport companies throughout the process of manufacturing and delivering each product to the customer. Various issues such as uncertainty in predicting demand, understock, overstock, and logistics and transport delays may interfere with the operation of the supply chain and cause uncertainty and complexity when its participants are not perfectly matched [2–4]. The concept of smart supply chains has been proposed to reduce variations in supply and demand occurring due to supply chain complexity and uncertainty [5]. Smart supply chains attach various sensors and RFID tags to devices and products used in product manufacture and transport so relevant data can be collected and analyzed to operate and make decisions based on meaningful data [5, 6]. It is important for the technical capabilities of these smart supply chains to be able to respond efficiently to insufficient resources. They can increase efficiency using smart technology, but only if the necessary internal core competences for decision-making and for capabilities to respond to changes in the internal/external business environment are in place first. Therefore, leading and participating companies in smart supply chains must have dynamic capability in order to react flexibly to environmental changes and achieve good performance [7].

Previous research on smart supply chains has focused on the use of digitalized supply chains and studied, for example, the effects of information technology competence on supply chains; the effect of 4th industrial revolution technology on logistics performance; the effect of IoT technology on supply chains; the effect of 4th industrial revolution technology on supply chain performance; tracking logistics objects for smart SCM; basic factors in smart supply chain composition; research on smart SCM models using RFID; research on smart SCM models based on synchronization of logistics information; and key technologies and supply chain management in the 4th industrial revolution [5, 7–13]. Empirical research to analyze structural relationships between dynamic capability and business performance in a smart supply chain environment is therefore lacking, and studies in this field are needed.

The purpose of this study is therefore to empirically analyze the effects of environmental uncertainty and dynamic capability on business performance in a smart supply chain environment. Through this study uncertain business environments and proposes goals to achieve ongoing competitive advantages by confirming structural relationships connecting the acquisition of dynamic capability to business performance.

2   Theoretical Background and Hypotheses

2.1   Smart Supply Chains

The huge wave of the 4th industrial revolution amplifying uncertainty as well as changes in technology and markets has profound effects on companies of all sizes [14]. As technology develops, intelligent supply chain management technology is being used to strengthen competitiveness and ensure corporate productivity by improving logistics information [8]. Smart supply chains are systems that fuse IoT and cyber-physical systems with a flow of data and products with various sensors or RFID tags attached to, for example, devices or products to solve problems of dynamic complexity when operating the chains. Large volumes of information previously unavailable in supply chains are provided throughout the entire process so the operator can use them to operate and make decisions. This reduces inefficiency and enables flexible responses to business environments to achieve results [5]. Technological development is accelerating changes to smart supply chains in order to respond to uncertainty. If changes in large volumes of data generated between traders within the smart supply chain are collected in real-time and used effectively, they can alleviate various issues such as supply–demand mismatch, overstock, understock, and transport delays [5].

2.2   Dynamic Capabilities

Dynamic capabilities consist of information and knowledge acquired or learned to control, integrate, or rearrange resources in order to achieve competitive advantages while rapidly responding to a quickly changing business environment from an extensive viewpoint of core competences, the foundation of resource-based theory. They can be called systematic or organizational capabilities [15]. Dynamic capabilities propose grounds for the need to strengthen and maintain capabilities and ensure competitive advantages in very uncertain business environments while strengthening the dynamics of needed internal capabilities [16]. The concept of dynamic capability varies between scholars with a range of opinions regarding its constituents [17–21].

Exploring market opportunities is an essential element of sustainable corporate growth, and if a company can locate an opening more quickly than its competitors, this will be a good driver for its profit and growth [22]. The act of detecting market environments and collecting data while pursuing growth is itself the process of exploring opportunities [23]. Capability in exploring opportunities is a company's action to detect opportunities and risks in the changing market environment and appears as a process of monitoring investment and technological changes to establish response strategies and gain market information based on those market changes as well as exchanging information about new opportunities [24]. From a smart supply chain perspective, the disruptive information technology of the 4th industrial revolution is created as a management system connected to the supply chain, and capability with this smart IT affects the dynamic capability to be able to detect, explore, and seize business opportunities and risk elements [16]. Therefore, in order for a company to continue to thrive, it is important to constantly explore business opportunities, and these capabilities may be accumulated through a focus on the process of pursuing opportunities itself [25].

The construction and acquisition of resources and capabilities to respond to the rapidly changing market environment and maintain ongoing competitive advantages is an important factor for companies. Penrose [26] divided resources into physical and human resources, and Wernerfelt [27] defined tangible and intangible assets belonging semi-permanently to companies. The construction and acquisition of resources and capabilities to respond to the rapidly changing market environment and maintain ongoing competitive advantages is an important factor for companies. The acquisition of external resources is an especially crucial factor for small and medium businesses due to their relatively insufficient resources and capabilities [22]. Furthermore, acquisition of external resources and their application to the changing business environment plays the role of leverage for the sake of a company's survival, enabling it to achieve better performance than its competitors. Resource acquisition capability, the ability to acquire new information and knowledge from outside the company due to insufficient resources in this changing environment, is a key factor in a company's dynamic capability [28].

For the sake of a company's survival and sustainable growth in a quickly changing market environment, reallocation of that company's resources is inevitable [22]. This is because companies' resources, capabilities, and structure must be integrated or reconfigured from mature business or declining fields of business to new business opportunities [24]. In other words, resource reconfiguration capabilities are a company's skill in integrating and reallocating previous and new resources to convert into new resources [19, 24]. This conversion and adjustment of resources is necessary for a company to survive and maintain its competitive advantages in a changing business environment. A company's skill in integrating and combining resources plays a key role in product innovation. The process of product innovation demands active combination of tangible and intangible assets by businesses and managers, and the skill to combine these is at the heart of dynamic capability regardless of a company's size [29]. However, since small and medium companies are weaker than large ones in terms of technology, capital, manpower, business, etc. [30], they may be limited in their ability to transform according to smart supply chain environment changes right away [31]. However, resource reconfiguration, a part of dynamic capability, is an important factor in creating new technology through readjusting and reallocating previously and newly acquired resources in order to respond to a changing environment, even if such resources are limited [32].

2.3 Dynamic Capabilities and Business Performance

Dynamic capabilities are playing a larger role in business performance while levels of uncertainty and competition increase in the business environment recently. Organizations must acquire dynamic capabilities for their own survival and performance improvement in a rapidly changing business environment. Dynamic capabilities are skills enabling a company to readjust its resources to achieve new competitive advantages, and such capabilities are strategic rather than temporary problem-solving abilities that show sustainability over time [18]. Companies must acquire enough organizational capital to create excellent results while responding to environmental changes and can only achieve sustained competitive advantages if they have the dynamic capability to use this capital according to their vision and mission. Furthermore, dynamic capabilities are controlled by past activities, current position, and future processes which are recreated in a new market environment [22]. Makadok [33] stated that the capability to explore opportunities, resource acquisition, and the capability to reconfigure resources, all elements of dynamic capability, show in a company's profits. Zahra et al. [34] proposed that dynamic capabilities indirectly affect business performance. Kim et al. [22] stated that elements of dynamic capability have significant effects on a company's financial and non-financial business performance. It has additionally been proposed that a company's dynamic capabilities have a positive effect on its financial or non-financial performance and are an element in its competitive advantages [35, 36]. This study therefore has the following hypotheses based on previous research.

H1. Dynamic capabilities have a positive effect on financial performance.
H2. Dynamic capabilities have a positive effect on non-financial performance.

2.4 Environmental Uncertainty

The business environment is changing at such a rapid speed that it is difficult for companies to predict by combining all elements related to direct/indirect business, and the number of factors businesses must take into consideration is gradually increasing as well [37]. Amidst this rapid change, environmental uncertainty from the point of view of companies and the supply chain is an inherent condition in interactions between companies, and this means that while new products and technologies are frequently released among fierce competition and customer demand strongly fluctuates, government policies often change and the results of decisions are difficult to accurately predict [38]. Ganbold and Matsui [39] subdivided environmental uncertainty into supply uncertainty, demand uncertainty, and technical uncertainty in a study on integration of supply chain management. In general, it can be predicted that the more complex the environment and the more dynamic the awareness, the more companies will strive to actively set targets and reduce that uncertainty [40]. Hong and Cheong [41] proposed that the speed of environmental change affects information sharing and cooperation, while Jo et al. [42] stated that improvement of business skills is an essential factor to achieve goals in an uncertain market environment and adapt to market changes. Furthermore, according to research by Ji and Pyun [43], individual companies with resource limitations aim to actively cope with environmental uncertainty to reduce it while accelerating product release and decision-making and improving their responsivity to market changes in order to use this environmental uncertainty as an opportunity. Lee [44] found that the greater the uncertainty, the greater the logistics performance in a study of differences in logistics performance according to environmental uncertainty. Therefore, the environmental uncertainty perceived by members of the supply chain can be expected to affect dynamic capability and business performance, and so we propose the following hypotheses.

H3. Environmental uncertainty has a regulative affect in the relationship between dynamic capability and financial performance.
H4. Environmental uncertainty has a regulative affect in the relationship between dynamic capability and non-financial performance.

3   Research Method

3.1   Data Collection

In A total of 290 surveys of employees of small and medium manufacturers understanding supply chains were distributed by mail from January to February of 2020 in order to analyze the structural relationship between dynamic capability and business performance in a smart supply chain environment. Of these, 219 were collected, and 8 were excluded due to unfaithful responses, leaving 211.

The manufacturers who were subjects of this study could be broadly classified into suppliers and purchasers, with 83 suppliers (39.3%) and 128 purchasers (60.7%). There were 159 companies less than 20 years old (75.4%) and 52 more than 20 years old (24.6%). In terms of number of employees, 205 companies had less than 100 (97.2%), while 6 (0.8%) had more than 100. There were 149 companies (70.6%) with less than 10 billion won in sales and 62 (29.4%) with more than 10 billion won in sales. Regarding partners, 180 companies (85.3%) had less than 50 partners, while 31 companies (14.7%) had more than 50 partners.

3.2   Measurement

The variables of dynamic capability and business performance used in this study were derived from previous research and measured on a 5-point Likert scale. Items measured for each variable were tested for validity and reliability in preceding research, and these were adjusted and used for this study (Table 1).

4 Empirical Analysis

4.1 Testing the Measurement Model

To test this study's hypotheses, we had to test the fitness, reliability, and validity of the measurement model, and we used confirmatory factor analysis for this. We judged fitness of the measurement model by using absolute fit index, incremental fit index, and parsimonious fit index. The results of confirmatory factor analysis of the measurement model showed that $\chi2 = 1709.131$ (P = 0.000) and $\chi2/df = 2.735$, below the recommended level of 3, and RMR = 0.054, below the recommended figure of 0.08. Furthermore, PGFI = 0.607, above the recommended value of 0.6; CFI = 0.849 and TLI = 0.830, slightly lower than the recommended figure of 0.9 but still satisfactory; and PNFI = 0.697 and PCFI = 0.755, above the recommended level of 0.6. The measurement model proposed in this study was thus deemed to be fit. The reliability and validity of the measurement tool were thereby proven as the measurement model was deemed fit. The convergent validity of the measurement tool must have a standardized regression coefficient greater than 0.5 and ideally greater than 0.7. Construct reliability (CR) must be above 0.7 and average variance extracted (AVE) must be above 0.5 to be considered fit. Furthermore, for discriminant validity, the latent factor's AVE must exceed the squared value of the correlation coefficient between constructs (Ø2).

4.2 Testing the Structural Model

With the fitness of this study's measurement model and reliability and validity of this model confirmed, we implemented analysis of the structural model to clarify whether the theoretical relationships of the research model were supported by the data. The results of testing the fitness of the structural model showed that $\chi2 = 1703.81$ (P = 0.000) and $\chi2/df = 2.734$, below the recommended level of 3, and RMR = 0.056, below the recommended figure of 0.08. Furthermore, PGFI = 0.612, above the recommended value of 0.6, and CFI = 0.847 and TLI = 0.830, slightly below the recommended level of 0.9 but still satisfactory. PNFI = 0.703 and PCFI = 0.763, above the recommended value of 0.6, meaning that the proposed structural model is fit.

With the structural model used in this study deemed fit, we tested our hypotheses by measuring path coefficients between the research variables. Results of testing the hypotheses showed that in the relationship between dynamic capability and financial

Table 1 Sample characteristics

Variable		Item measured	References
Dynamic capability	Opportunity exploration capability	Benchmarking excellence in the same industry Occasional checking and analysis of market changes Seeking ideas for new product development Ongoing collection of competitor information Exploring, collecting, and analyzing market opportunities	[17, 18, 45, 24, 46, 22]
	Resource acquisition capability	Fostering experts for new business Acquiring external knowledge for new product development Building partnerships with other companies to develop capabilities Systematizing internal knowledge from employee experiences Acquiring resources for new product development	
	Resource reconfiguration capability	Process reconfiguration to introduce new technology and knowledge Integration and adjusting of management teams to respond to environmental changes Resource reallocation to suit new environments Changing technical facility processes to respond to environmental changes Reconfiguration of resources needed to develop new products	
Business performance	Financial performance	Increased sales Increased operating profit Improved cash flow Increased market share	[22, 47, 48]
	Non-financial performance	Improving participants' capabilities and technical level Improving customer satisfaction Improving the company and product image Improving employee satisfaction	
Environmental uncertainty		Demand uncertainty Supply uncertainty Competitor uncertainty Technical uncertainty	[2, 4, 49]

performance, the coefficient value of opportunity exploration capabilities toward financial performance was (β) = 0.349 (C.R. = 3.016, p = 0.003), the coefficient value of resource acquisition capability toward financial performance was (β) = 0.064 (C.R. = 0.454, p = 0.650), and the coefficient value of resource reconfiguration capability toward financial performance was (β) = 0.223 (C.R. = 1.940, p = 0.052), and we therefore selected Hypothesis H1-1 and rejected Hypotheses H1-2 and H1-3.

Moreover, in the relationship between dynamic capability and non-financial performance, the coefficient value of opportunity exploration capability toward non-financial performance was (β) = 0.329 (C.R. = 3.535, p = 0.000), the coefficient value of resource acquisition capability toward non-financial performance was (β) = 0.265 (C.R. = 2.303, p = 0.021), and the coefficient value of resource reconfiguration capability toward non-financial performance was (β) = 0.276 (C.R. = 2.966, p = 0.003), and we therefore selected all three Hypotheses H2-1, H2-2, and H2-3.

4.3 Moderating Effects of Environmental Uncertainty

Results of analyzing the moderating effects of environmental uncertainty in the relationship between dynamic capability and business performance showed a correlation between opportunity exploration capability and non-financial performance (t = $-$2.026, p = 0.044) and a correlation between resource reconfiguration capability and non-financial performance (t = -2.850, p = 0.005), suggesting that environmental uncertainty has a significant regulating effect.

5 Conclusions

For this study, we reviewed the role of variables through literature research and empirically tested their influence relationships to determine the effect of a company's dynamic capabilities on its business performance so small and medium manufacturers can respond to rapidly changing business environments within a smart supply chain.

Setting hypotheses regarding the interaction between dynamic capability and business performance showed that dynamic capabilities had a partially significant effect on business performance. In more detail, opportunity exploration capabilities affected both financial and non-financial performance, and resource acquisition capabilities and resource reconfiguration capabilities only affected non-financial performance. These results support the conclusions of Eisenhardt and Martin [18], Kim and Huh [36], and Kim et al. [22] and confirm that small and medium companies' dynamic capabilities are positively correlated with their business performance. As small and medium companies are relatively lacking in resources and struggle to sustain competitive advantages in a rapidly changing, dynamic environment, they need dynamic capabilities to respond to these constantly changing situations. However, resource acquisition capabilities and resource reconfiguration capabilities were not found to have a significant effect on financial performance. We can infer from this that since financial performance is a short-term, direct measure of a company's past business activities, there is little direct, short-term correlation between resource acquisition capabilities and resource reconfiguration capabilities which appropriately reallocate to a company's business capabilities and act as leverage.

Moreover, the results of testing the regulating effects of environmental uncertainty in the relationship between dynamic capabilities and business performance showed that environmental uncertainty does have a regulating effect only in the relationship between dynamic capability and non-financial performance. This shows that environmental uncertainty has a regulating effect in the more long-term results appearing in the relationship with non-financial performance rather than in the short-term results of financial performance, and this could imply a need to establish strategies early in order to improve long-term responsiveness to environmental uncertainty.

This study deviated from previous research which focused on the role of IT on supply chain performance and logistics performance in a smart supply chain environment for the sake of small and medium companies' business performance to confirm dynamic capability factors that companies must improve to maintain competitive advantages in an uncertain business environment and to establish a structural relationship linking acquisition of dynamic capabilities to business performance.

For this study, we set our primary subjects as small and medium manufacturers to establish a structural relationship between dynamic capabilities and business performance in a smart supply chain environment. However, the business conditions of these companies may differ, and we did not measure or categorize them systematically according to their stage of smart supply chain construction but rather generalized them all. More meaningful studies are expected in the future if differences in awareness of dynamic capabilities is confirmed by classifying clear differences in the characteristics of different industries and the roles of suppliers and purchasers as we selected suppliers and purchasers in the supply chain environment based on their subjective responses.

References

1. Kim JH (2016) In the era of the 4th industrial revolution, seeking strategic responses to future social changes. R&D InI 15:45–58
2. Kim SO, Youn SH (2008) A study on the effects of environment uncertainty and inter-firm collaboration practice on supply chain flexibility. Korean Assoc Indust Bus Administ 23(1):337–364
3. Lee GD (2013) An empirical study on the relationships among environmental uncertainty, management by objectives and organization satisfaction. Tax Account Res 22(1):179–202
4. Kim JY, Bang HY (2014) The effects of output sector uncertainty on dependence, commitment and strategic performance: a comparative analysis on korean and american manufacturers. Korea Assoc Int Commer Inf 16(1):163–183
5. Shin JC, Lim OK, Park YH, Song SH (2017) A study on determining priorities of basic factors for implementing smart supply chain. J Korean Soc Supply Chain Manage 17(1):1–12

6. Lifang W, Yue X, Alan J, Yen DC (2016) Smart supply chain management: a review and implications for future research. Int J Logist Manage 27(2):395–417
7. Moon TS, Kang SB (2014) An empirical study on the impact of it competence on supply chain performance through supply chain dynamic capabilities. Korean Manage Rev 43(1):245–272
8. Kim JG (2013) A study on the logistics information synchronization based smart SCM model. KIPS Trans Softw Data Eng:311–318
9. Lee KB, Baek DH, Kim DH (2016) A study on the effect of the IoT technology on SCM. J Inf Technol Serv 15(1):227–243
10. Lee CB, Noh JH, Kim JH (2017) A study on the perception of the impact of fourth industrial revolution on the performance of logistics management. Korea Logist Rev 27(5):1–12
11. Chung CC (2017) The fourth industrial revolution—an exploratory study on main technologies and supply chain management. Korea Logist Rev 27(6):193–209
12. Noh JH, Lee KN (2018) The effect of supply chain related 4th industrial revolution technology on BSC supply chain performance. Korea Logist Rev 28(5):53–64
13. Kwak KG, Hwang SY, Shin DJ, Park KW, Kim JJ, Park JM (2020) Study of logistics object tracking service for smart SCM. J Korean Inst Indust Eng 46(1):78–81
14. Kim HT, Kwon SJ (2019) Exploring the industrial structure and innovation policy for sustainable growth of machinery industry in South Korea. Sci Technol Policy Inst 2(1):107–131
15. Huh YH, Lee C (2012) Determinants of dynamic capability and its relationships with competitive advantage and performance in foreign markets. Int Bus J 23(1):33–73
16. Rhee YP (2020) The relationships between IT capability, dynamic capability and international performance in Korean SMEs. Int Bus Rev 24(1):107–120
17. Teece DJ, Pisano G, Shuen A (1997) Dynamic capabilities and strategic management. Strateg Manage J 18(7):509–533
18. Eisenhardt KM, Martin JA (2000) Dynamic capabilities: what are they? Strateg Manage J 21(10):1105–1121
19. Bowman C, Ambrosini V (2003) How the resource-based and the dynamic capability views of the firm inform corporate-level strategy. Br J Manage 14(4):289–303
20. Wang C, Ahmed PK (2007) Dynamic capabilities: a review and research agenda. Int J Manage Rev 9(1):31–51
21. Denford JS (2013) Building knowledge: developing a knowledge-based dynamic capabilities typology. J Knowl Manage 17(2):175–194
22. Kim JK, Yang HC, Ahn TD (2017) The moderating effect of manufacturing process type on the relationship of dynamic capabilities and business performance of SMEs. Asia-Pacific J Multimedia Serv Converg Art Human Sociol 7(8):141–151
23. Helfat C, Finkelstein S, Mitchell W, Peteraf M, Singh H, Teece D, Winter S (2007) Dynamic capabilities: understanding strategic change in organizations, pp 1–18
24. Teece DJ (2007) Explicating dynamic capabilities: the nature and microfoundations of (sustainable) enterprise performance. Strateg Manage J 28(13):1319–1350
25. Ambrosini V, Bowman C (2009) What are dynamic capabilities and are they a useful construct in strategic management? Int J Manage Rev 11(1):29–49
26. Penrose ET (1959) The theory of the growth of the firm. Oxford
27. Wernerfelt B (1984) A resource-based view of the firm. Strateg Manage J 5(2):171–180
28. Verona G, Ravasi D (2003) Unbundling dynamic capabilities: an exploratory study of continuous product innovation. Indust Corpor Change 12(3):577–606
29. Augier M, Teece DJ (2009) Dynamic capabilities and the role of managers in business strategy and economic performance. Organ Sci 20(2):410–421
30. Kim YY, Park YS (2017) Fourth industrial revolution and SME supporting policy 20(2):387–405
31. Park CK, Kim CB (2019) Effect of agility capabilities for smart manufacturing and smart SCM implication on enterprise and supply chain performance in the 4th industrial revolution. J Small Bus Innov 22(4):23–67
32. Morgan NA (2012) Marketing and business performance. J Acad Mark Sci 40(1):102–119

33. Makadok R (2001) Toward a synthesis of the resource-based and dynamic capability views of rent creation. Strateg Manage J 22:387–401
34. Zahra SA, Sapienza HJ, Davidsson P (2006) Entrepreneurship and dynamic capabilities: a review, model and research agenda. J Manage Stud 43(4):917–955
35. Hwang KY, Sung EH (2015) The relationships between dynamic capabilities, innovation performance and performance of export venture firms. J Int Trade Commer 11:401–420
36. Kim GT, Huh MG (2016) Dynamic capabilities and competitive advantages: the moderating effect of environmental dynamism. J Strat Manage 19(3):81–103
37. Huh SJ, Lee JY, Hyeon JW, Choi YS (2018) Business environment uncertainty and real earnings management. Korean Corpor Manage Rev 25(6):77–101
38. Miller D (1988) Relating Porter's business strategies to environment and structure: analysis ans performance implications. Acad Manage J 31(2):280–308
39. Ganbold O, Matsui Y (2017) Impact of environmental uncertainty on supply chain integration: empirical evidence. J Jpns Oper Manage Strat 7(1):37–56
40. Park SM, Park CW (2018) A study on the effect of multi-dimensionality of environmental uncertainty on smart SCM factors and corporate management performance. Korea J Logist 26(2):105–126
41. Hong KS, Cheong KW (2004) The impact of environmental clockspeed, information sharing, and collaboration on supply chain performance. Korean Small Bus Rev 26(2):77–100
42. Jo YG, Lee HG, Ha KT (2007) The relationship among the utilization of IT based on realizing RTE, agility capabilities and company performance. Entrue J Inf Technol 6(2):113–127
43. Ji SG, Pyun HS (2009) The relationship among environmental uncertainty, marketing agility, marketing performance. Korea J Bus Administ 22(2):1013–1035
44. Lee CS (2012) The moderating effects of logistics management system sophistication and strategy on the relationship of environmental uncertainty and logistics performance. Korea Logist Rev 22(1):183–209
45. Zollo M, Winter S (2002) Deliberate learning and the evolution of dynamic capabilities. Organ Sci 13:339–351
46. Chun JI, Lee BH (2016) The effect of dynamic capabilities on international performance of Korean exporting SMEs: the moderating role of environmental dynamism and firm type. Int Bus Rev 20(1):45–74
47. Karplan RS, Norton DP (1966) Using the balanced scorecard as a strategic management system. Harvard Bus Rev:75–85
48. Lee JD, Lee YB, Bae YS (2014) The effects of SMEs' core competency and competitive strategy on their business performance. J Korean Entrepreneurs Soc 9(3):154–183
49. Walker G, Weber D (1984) A transaction cost approach to make-or-buy decision. Adm Sci Q 29(3):374–391

Analysis of the Relative Importance of Coaching Service Quality

Jinyoung Yang, Hyoboon Wang, and Donghyuk Jo

Abstract This study aims to analyze the relative importance between determinants and factors of coaching service quality. For this purpose, the hierarchical factors of coaching service quality are drawn by reviewing literature on coaching and using Delphi technique, and survey was conducted on expert coaches in the coaching field to analyze the relative importance of coaching service quality. As a result of analysis, the relative important of primary hierarchical factors of coaching service quality factors were Reliability, Expertise, Empathy, Responsive-ness, Effective-ness, and Tangibles from highest to lowest. In addition, the result of combined importance on secondary hierarchical factors showed Listening Actively, Positive Feedback, and Ethical Practice as important factors. Such results may be interpreted as establishing foundation of coaching relationship and forming close confidential relationships are important in improving coaching effects. Result of this study may be used as basic data in establishing an academic knowledge system of coaching and expanding the base for coaching.

Keywords Coaching · Coaching competency · Coaching service quality · AHP

1 Introduction

The modern era, which began at the end of the nineteenth century, brought upon signifi-cant socioeconomic changes unlike the previous society due to the effects of logical-rational thinking, and scientific positivism, and the world following it

J. Yang · H. Wang
Soongsil University, Seoul, South Korea
e-mail: yangyanggood@naver.com

H. Wang
e-mail: ynk8904@naver.com

D. Jo (✉)
Department of Business Administration, Soongsil University, Seoul, South Korea
e-mail: joe@ssu.ac.kr

is characterized as two significant changes as the beginning of postmodern era in the mid-twentieth century, which one of them is the transformation from industrial economy to service-based economy, and another one is the appearance of coaching industry in more complicated and growing social changes. Whereas the focus of economic activity in the modern era was the production of goods and money, it shifted to the production of nonmaterial convenience based on knowledge, information, and technology in the postmodern era. Also, people became aware through World War I and II that the rationality and reason of humans in the existing era don't just bring positive results, and unlike the modern era where an absolute standard or truth existed and laborers were considered production unit under hierarchical order and control, postmodern era formed a service-oriented society by acknowledging individual's diversity and changing the nature and idea of labor through reintegration of life and work [1–3].

Second important change in postmodern era is the appearance of coaching. Socioeconomic change in this era change the style and pattern of life, and brought significant changes in new perspective and value. Individual value became important in complex globalization, and human's consciousness gradually grew as the quality of life improved due to the increase of economic wealth. Growth of human's consciousness lead to the desire for respect and recognition beyond survival, and to the interest in social success, and people began to pay attention to self-improvement and growth as the meaning of life, happiness and desire for self-realization grew larger [4]. As a result, coaching focusing on growth and development came to the fore, and is being used in different areas including management, life, career, business, Christianity, relationship, etc. [5].

Service is a tangible and intangible substance that satisfies consumer's desire for expectation, and may be understood as a series efforts to satisfy clients [6]. It also evaluates and manages service quality in various service areas in order to satisfy client's expectation and draw satisfaction. Studies related to service quality is particularly active in medicine, and hospital management competitiveness is increased by acquiring positive effects in client loyalty and word of mouth effects through improved reuse intention beyond client satisfaction by accurately understanding, improving, and developing client's demand and necessity [7]. On the other hand, in coaching, objective evaluations are not made how the expectation of clients who received coaching service is satisfied. Mostly, client feedback is received on coaching service after agreed coaching sessions, but they mostly consist of subjective evaluations such as what they liked or self-insight, so there are still insufficient detailed and objective data on the effects of coaching and client satisfaction.

Therefore, this study intends to establish and develop service quality of coaching from service perspective based on coaching competency, and analyze the relative importance of coaching service quality using AHP analysis. The purpose of this pro-cess is to contribute to expanding base for coaching through the improvement of coaching quality by measuring and managing service quality, positive effects on client loyalty beyond client satisfaction, and establishment of quality factors of coaching service.

2 Theoretical Background

2.1 Coaching

The term coaching originated from a Hungarian four-wheeled carriage in the sixteenth century called kocsi, and refers to a personalized transportation service that departs from passenger's current position and arrives at the desired destination [8]. Since coaching is customized to coach, passenger, and situation, the definition of coaching has been suggested differently by scholars of different areas based on its nature and object, combined with academic ground theory. Based on various perspectives of scholars, definition of coaching can be categorized into the aspect that emphasizes promotion of study, aspect that emphasizes leadership, and aspect that emphasizes communication [9].

First is the definition of coaching that supports and promotes learning [10]. In this aspect, coaching is not teaching, but awakening of individual's potentials to help one study by oneself. In other words, it is a system that supports client's self-realization through a systematic and cooperative process based on problem-solving through individual's growth and self-directed learning experience [11]. Second is the definition of coaching approaching as a leadership type. In this perspective, coaching promotes client's behavioral change through conversation and discussion, to empower for further achievement [12]. Last is the definition of coaching as a communication skill. In this approach, coaching is a two-way communication skill based on respect for client, paying attention client's desire and promoting client's voluntary behavior, and is a communication process that enables continuous development of competence [13].

In addition, International Coach Federation (ICF) defines coaching as a professional relationship which helps achieve extraordinary results in life, career, business, or organization, and Korea Coach Association (KCA) defines coaching as a horizontal partnership that helps realize the highest values by maximizing individual's and organization's potentials [14, 15]. Like this, definitions of coaching varies based on perspective, but the common key elements of coaching may be summarized as a partnership that supports an individual and an organization to demonstrate one's potentials to grow and develop, and a two-way communication system [16].

Coaching is different from similar areas such as counseling, education, training, consulting, and mentoring in the fact that it offers a horizontal partnership, two-way communication, client as the subject of problem-solving, growth and development, client's voluntary and proactive participation, minimal intervention of an expert, and active delegation of authority to a client [17].

2.2 Coaching Competency

Generally, competency is divided into an observable overt element that can successfully perform a certain task and an unobservable inherent characteristic [18], and coaching competency refers to the intrinsic and extrinsic characteristics of a coach including values, attitude, knowledge, and techniques, as a whole [19].

In November 2019, ICF categorized core competencies of a professional coach into 8 characteristics. The 8 core competencies include demonstrating ethical practice, embodying a coaching mindset, establishing and maintaining agreements, cultivating trust and safety, maintaining presence, listening actively, evoking awareness, and facilitating client growth, which are supported with 63 sub-guidelines to provide the standard of competencies as a professional coach [14]. In Korea, KCA developed a standard coaching competency model for Korean coaches in 2019 through experts' validity verification. It divides the competency into 'Being a coach' and 'Coaching', where 'Being a coach' consists of the ability to 'realize philosophy of coaching', coach's 'self-management' ability, ability to 'detect' referring to coach's awareness and re-flection using intuition, and coach's 'expertise' referring to accumulation and use of knowledge and experience for coaching operation. 'Coaching' competency consists of 'participation' ability referring to a horizontal partnership participating in client's changes and growth, 'expansion of consciousness' ability to help client's change in perspective and expansion of awareness, 'communication' ability enhancing coaching effects such as effective questions and active listening, and 'process management' ability to help client achieve goal [15].

Stowell [12] considered establishment of partnership, cooperation, interest in demand, empathy, compliment, patience, and creation of environment for interaction as important factors of coaching competency, and Law [20] considered empathy, kind personality, patience for ambiguity, awareness, diplomatic ability, facilitation skill, learning skill, interview skill, listening skill, and change management as factors of coaching competency. Song [21] coaching ethics, listening skill, questioning skill, acknowledging skill, feedback skill, summarizing skill, process application skill, skill to use coaching diagnostic tool, time management skill, business insight, understanding of organizational issue, strategic thinking, coaching performance maintenance and management, and ability use experience as important competencies. Do and Kim [22] deduced autonomy, optimism, hope, relationship, and listening to question and feedback as components of coaching competency.

2.3 Coaching Service Quality

As industrial economy transformed into service-based economy, the importance of service began to rise, and the service society, formed as a result of huge socioeconomic changes, shifted from provider-centered to consumer-centered and grew to-wards client-oriented perspective.

Service quality emphasizes subjective judgment based on client's individuality and diversity and is difficult to measure objectively, therefore, it needs to be evaluated through indirect clues. In other words, it is measured by subjective expectation and perception of a client who experienced the service. In terms of various previous studies on service quality, Grönroos [23] called it the result of comparison evaluation process client's perceived service and expected service, Parasuraman et al. [24] explained service quality as the degree and direction of difference between expected service and perceived service, and defined service quality as client's overall judgment or attitude towards the excellence of service. Cronin and Talor [25] raised an objection to Parasuraman et al. [24] expectation and performance management before and after client's service experience, and insisted that it needs to be measured only through the perceived service performance, eliminating the expectation on service. After, studies have been conducted in the 2000s, and Shen et al. [26] defined it as the degree of satisfaction beyond client's expectation, and stated that service quality can be measured through client satisfaction. Study by Etgar and Fuchs [27] explained that it connotes psychological decision on service perceived by client's expectation.

As interest and researches on service quality became active, attempts to establish and measure quality factors have been made. The model most frequently used in studies is SERVQUAL model by Parasuraman et al. Their early study suggested 10 quality factors perceived by clients who experienced service, but were criticized that service expectation performance measurement based on experience may be difficult to apply in actual field, and revised them into 5 factors in the follow-up study, including tangibles, reliability, responsiveness, assurance, and empathy. SERVQUAL model is usefully utilized to measure and manage service quality in various service areas, and contributes significantly to management performance by studying client's desire, need, and expectations, improve client satisfaction, and secure loyal clients; it is widely used in studies of different areas from management to hotel, education, medicine, beauty, and counseling.

Coaching is a horizontal conversation with a process, which is provided to help achieve client's desired goal and future-oriented growth. Clients receiving coaching can establish detailed implementation plans for goals and realize voluntary actions by introspect themselves again and discover possibility and potential in themselves. Performance of coaching may be evaluated based on 'how much client changed or grew', and coaching may be considered successful when the client who received coaching seeks a new life to achieve the goal through different thinking and behavior. For successful coaching, coaches must possess coaching competencies to interact with clients and implement in the coaching scene, and success of coaching can be expected when they properly possess and realize such competencies. Eventually, the quality of coaching needs to be improved in order to bring benefits and satisfaction to the client and improve coaching performance. In other words, coaching competency can improve the quality of coaching service and enhance customer satisfaction, therefore, coaching competency-based service quality needs to be developed, and coaching as an area of service industry needs studies that establish, measure, and evaluate quality factors of coaching service using the service quality measurement model.

Fig. 1 Coaching service quality research methods and procedures

3 Research Design

3.1 Research Method

To evaluate the quality factors of coaching service and relative important, the study was designed and implemented in 3 steps. First step includes literature review and research of previous study data, and analyzing core coaching competencies by ICF and coaching competencies by KCA to draw primary hierarchical factors of coaching competency-based coaching service quality. Second step used Delphi technique. Delphi survey was collected through 2 professors of business administration and 13 coaching experts with doctorate related to coaching to draw secondary hierarchical factors of coaching service quality. Third step used AHP techniques to analyze the relative importance between hierarchies. Based on final hierarchy structure deduced through Delphi technique, a 9-point scale survey was written on the relative importance of evaluation factors related to coaching service quality.

AHP technique was developed by Thomas L. Satty, and is still widely used and studied in decision-making area; the biggest strength of AHP technique is the reflection of qualitative and quantitative evaluations on various alternatives. Service quality for clients is intangible and contains psychological nature, and using AHP technique can effectively analyze the psychological area that reflects individual's subjective thinking, emotion, and belief (Fig. 1).

3.2 Study Model

For this study, the SERVQUAL model, data from previous studies, opinions from coaching experts, and Delphi questionnaire were comprehensively reflected and developed as shown in Table 1.

Table 1 Hierarchical structure of coaching service quality

Coaching service quality	Primary hierarchical factors	Secondary hierarchical factors
	Expertise	Coaching knowledge and experience
		Listening actively
		Effective question
		Intuition and Insight
	Tangibles	Coach's attire and appearance
		Proper place and atmosphere for coaching
		Proper coaching tool
		Coaching style (facing/non-facing)
	Effectiveness	Attitude change towards future
		Promotion of self-directed learning
		Self-understanding and discovery of strength
		Voluntary goal setting and execution
	Reliability	Ethical practice
		Agreement and maintenance of coaching relationships
		Consistent sincerity
		Forming and maintaining close confidential relationship
	Empathy	Respect for client's unique individuality
		Curiosity and focus on client
		Positive attitude
		Genuine and warm attitude
	Responsiveness	Flexible application of various coaching methods
		Positive feedback
		Pacing such as breathing, speed and tone
		Exploring the other side and reflecting meaning

Table 2 Demographic characteristics of coaching expert

Items		Sample size	Ratio (%)
Age group	Under 20 s	0	0
	30 s	1	4.0
	40 s	10	40
	50 s	11	44
	60 s	3	12
Coaching field	Business coaching	3	12
	Life coaching	17	68
	Career coaching	1	4.0
	Learning coaching	1	4.0
	Etc	3	12
Coaching career	Less than 3 years	4	16
	3 years or more and less than 6 years	9	36
	6 years or more and less than 9 years	5	20
	9 years or more and less than 12 years	6	24
	12 years or more	1	4.0
Total		25	100.0

3.3   Measurement Tool

This study used AHP technique to deduce determinants and relative importance evaluation of coaching service quality, and designed the survey as paired comparison as seen in Table 1. In order to secure accuracy of survey, the operational definition of primary hierarchical factors of coaching service quality and example of measurement were presented in Table 2. Survey questions were designed as paired comparison on 9-point scale, which is used in AHP technique, to calculate the importance.

4   Analysis and Results

4.1   Data Collection and Inspection

This study conducted survey from June 10 to June 20, 2020 (10 days) on 36 experts who acquired professional coach certificate accredited by KCA (Korea Coach Association) and are working as a coach. Excluding 11 copies containing errors in the analysis process, 25 respondents were selected.

Reliability of AHP technique is considered secured when the consistency ratio (C.R) distinguishing error of respondent's decision is 0.1 or below [32]. Among 36 collected samples, responses with C.R over 0.1 (11 copies) were eliminated, and

Table 3 Consistency index by hierarchical factors

	Primary hierarchical factors	Secondary hierarchical factors					
		Expertise	Tangibles	Effectiveness	Reliability	Empathy	Responsiveness
C.R.	0.0172	0.0066	0.0283	0.001	0.0195	0.0063	0.0152

anal-ysis was performed on final 25 copies with secured reliability, in which the verifica-tion result of C.R values of each hierarchical factor is displayed in Table 3.

Relative Importance Analysis Result Analysis result of relative importance of coach-ing service quality is displayed in Table 4.

Based on analysis result of primary hierarchical factors, reliability (0.247) is the most important factor, followed by expertise (0.202), empathy (0.200), responsive-ness (0.168), effectiveness (0.140), and tangibles (0.043). In terms of the importance of secondary hierarchical factors, the first factor of expertise is ability to listen active-ly (0.443), followed by effective question (0.225), intuition and insight (0.216), and coaching knowledge and experience (0.116). In terms of tangibles, proper place and atmosphere for coaching (0.345) is the most important factor, followed by proper coaching tool (0.274), coach's attire and appearance (0.214), and coaching style (0.167). Importance of effectiveness factors includes self-understanding and discov-ery of strength (0.390), followed by voluntary goal setting and execution (0.256), atti-tude change towards future (0.204), and promotion of self-directed learning (0.151). Importance of reliability includes ethical practice (0.270), followed by form-ing and maintaining close confidential relationship (0.258), consistent sincerity (0.249), and agreement and maintenance of coaching relationship (0.223). Im-portance of empathy factors includes genuine and warm attitude (0.283), followed by curiosity and focus on client (0.263), respect for client's unique individuality (0.243), and positive attitude (0.212). Lastly, importance of responsiveness factors includes posi-tive feedback (0.410), followed by exploring the other side and reflecting meaning (0.250), pacing such as breathing, speed and tone (0.206), and flexible ap-plication of various coaching methods (0.135).

5   Conclusions

Future society in the 4th Industrial Revolution Era will bring huge socioeconomic changes based on AI, robotic technology, and biotechnology, and there will be con-siderable changes in the form and properties of service [28]. Service economization is a global trend, 70% of jobs are created in service industry, and the importance of service industry is more significant than other industries in the 4th Industrial Revolu-tion Era that the interest and demand in service are expected to grow further [29]. Coaching allows clients to have an insight and think through effective questions in systematic conversations with processes. In other words, coaching is a very im-portant industry in reinforcing competencies that humanity needs to possess in future

Table 4 The relative important of coaching service quality factors

Primary hierarchical factors			Secondary hierarchical factors			Combined importance	
Factors	Importance	Ranking	Factors	Importance	Ranking	Importance	Ranking
Expertise	0.202	2	Coaching knowledge and experience	0.116	4	0.0234	18
			Listening actively	0.443	1	0.0894	1
			Effective question	0.225	2	0.0454	11
			Intuition and insight	0.216	3	0.0436	12
Tangibles	0.043	6	Coach's attire and appearance	0.214	3	0.0092	23
			Proper place and atmosphere for coaching	0.345	1	0.0148	21
			Proper coaching tool	0.274	2	0.0117	22
			Coaching style (facing/non-facing)	0.167	4	0.0071	24
Effectiveness	0.140	5	Attitude change towards future	0.204	3	0.0667	17
			Promotion of self-directed learning	0.151	4	0.0550	20
			Self-understanding and discovery of strength	0.390	1	0.0615	8
			Voluntary goal setting and execution	0.256	2	0.0637	15
Reliability	0.247	1	Ethical practice	0.270	1	0.0285	3
			Agreement and maintenance of coaching relationships	0.223	4	0.0211	7
			Consistent sincerity	0.249	3	0.0546	5

(continued)

Table 4 (continued)

Primary hierarchical factors			Secondary hierarchical factors			Combined importance	
Empathy	0.200	3	Forming and maintaining close confidential relationship	0.258	2	0.0358	4
			Respect for client's unique individuality	0.243	3	0.0486	10
			Curiosity and focus on client	0.263	2	0.0526	9
			Positive attitude	0.212	4	0.0424	13
			Genuine and warm attitude	0.283	1	0.0566	6
Responsiveness	0.168	4	Flexible application of various coaching methods	0.135	4	0.0228	19
			Positive feedback	0.410	1	0.0693	2
			Pacing such as breathing, speed and tone	0.206	3	0.0348	16
			Exploring the other side and reflecting meaning	0.250	2	0.0422	14

society and adapt to future changes. Therefore, acknowledges the necessity to ex-pand the base of coaching, therefore, develop the model of service quality determi-nants in coaching service and analyze relative importance.

In the primary hierarchical factors, reliability (0.247) appeared as the most im-portant factor. This is because the coaching relationship and effect of coaching var-ies based on the formation of confidence with client. In counseling, formation of relation with client is called rapport, and rapport is client's feeling towards trust or immersion to maintain relationship with the client. Rapport building induces positive emotions in client and contributes to client satisfaction that it is used in mar-keting as well. Coaching prioritizes in confidential relationship with client for an ef-fective coaching session. Passmore [30] suggested 5 tasks for professional coaching activ-ities, in which the third task is establishing a consensual code of conduct. Prior to coaching, coach must participate in the consensual coaching relationship with client and process with consistent sincerity and responsibility. Second factor in pri-mary hierarchical factors of coaching service quality is expertise. Expertise is the skill or competency that is the most importantly used in the interaction between coach and client in coaching scene, which can be typically identified as active listening and effective question. Previous studies on coaching competency include listening and question in most competencies. Also, 11 core coaching competencies sug-gested by ICF include active listening and powerful questioning as a part of effective communi-cation skills. Through active listening, coaches listen to client's nonverbal emotions, suggestions, perception, value, interest, and belief, and allows client to be-come fully aware of and express oneself by delivering the intended intention or key point. Based on this, coaches asks effective questions so clients can pursue the de-sired target, and client can have new and creative ideas, which can become the moti-vation of growth and development. In this coaching process, coaches provide proper positive feedback to client, and secondary hierarchical factor of responsiveness showed posi-tive feedback (0.410) as the most important factor, as well as general importance showing it as second important. Bandura [31] defines positive feedback as a catalyst for successful performance of action for a specific goal. Therefore, positive feedback such as recognition, support, and compliment in the conversation through listening and question in coaching service helps client's growth, and is very important in coaching service quality.

Empathy factor (0.200) in primary hierarchy was the third important factor, and in terms of sub-factors of empathy, sincere and warm attitude (0.283) appeared the most important quality factor in coaching. DiDonato and Krueger [32] considers sincerity as receptive ability and interprets it as dynamic status, not fixed, and Wick-ham [33] states that people don't become defensive and form trust when they recog-nize their partner as sincere, and both parties can be connected stronger. Parasura-man et al. [34] describes about inseparability in service quality, and defines that ser-vice does not separate purchasing behavior and consumption behavior, but produc-tion and consumption occur simultaneously. Therefore, coaching with service nature includes sufficient connection and interaction with client through coach's sincere attitude, which in turn helps with client's change and growth. Next, effectiveness (0.140) in primary hierarchy is a consequential quality, which consists of criteria to recognize client's change after coaching session, and includes 5 factors. Importance is relatively lower than other quality factors, which can be interpreted as the confi-dence that reliability, expertise, and empathy will positively lead to effectiveness of coaching. In terms of importance of sub-factors of effectiveness, self-understanding and discovery of strength (0.390) is the most important. Coaching stimulates client's thinking and brings new perspective and insight, so client discovers inner possibility and potential, which accelerates client's spontaneity and practice of self-directed life to set and execute the goal. Therefore, coaching experts acknowledge self-understanding and discovery of strength as the most important factor of the effec-tiveness of coaching.

If effects of postmodern era developed service industry, the upcoming 4th Indus-trial Revolution Era is emphasizing the various forms and importance of service. In addition, development and expansion of base of coaching industry is very important to develop creative talent as humans are at the center of service, and quality of coaching service needs to be improved as it can bring values to clients. Result of this study was obtained from professional coaches, and is limited in interpreting client's expectation and opinion on coaching. Therefore, it should be expanded to clients to measure and evaluate the coaching service quality. As one suggestion, coaching knowledge and experience in sub-factors of expertise was the second-lowest in the secondary hierarchy, and the 18th in overall importance. Coaching, with inseparable characteristics, includes dynamic interaction that listening, questioning, and intuition are more important that coaching knowledge and experience. However, there is no academic system established for coaching yet. The first of 5 challenges suggested by Passmore [30] is 'explaining the system of coaching knowledge'. In order to realize the identity and philosophy of coaching and establish it as a systemic field of study, its knowledge system must be established first. Therefore, preparation for the 4th Industrial Revolution Era needs to be made by agreeing with the importance of coaching knowledge and experience and establishing system as a field of study, and expand the use of coaching to contribute to organizations and society.

References

1. Habermas J (1981) Modernity versus postmodernity. New German Critique 22:3–14
2. Choi I-S (2015) Critique on western-centrism inherent in post-modernism: focused on its historical view. Korean Soc Polit Thought 21(2):99–121
3. McKnight S (2010) Spirituality in a postmodern age. Stone-Campbell J 13(2):211–224
4. Kim E-J (2018) The application effects of a coaching program on preservice childcare teachers. J Korean Coach Res 11(3):73–91
5. Hamlin RG, Ellinger AD, Beattie RS (2009) Toward a process of coaching? A definitional examination of "coaching", "organization development", and "human resource development. Int J Evid Based Coach Mentor 7(1):13–38
6. Olorunniwo F, Hsu M, Udo G (2006) Service quality, customer satisfaction and behavioral intentions in the service factory. J Serv Market 20(1):59–72
7. Asnawi A, Awang Z, Afthanorhan A, Mohamad M, Karim F (2019) The influence of hospital image and service quality on patients' satisfaction and loyalty. Manage Sci Lett 9:911–920
8. Berg ME, Karlson JT (2007) Mental models in project management coaching. Eng Manage J 19(3):3–13

9. Nam SE, Ryu K (2017) An action research study of development and implementation of coaching program for college students on academic probation (action research). Korean J Gener Educ 11(1):281–311
10. Edwards L (2003) Coaching the latest buzzword or a truly effective management tool? Ind Commerc Train 35(7):298–300
11. Whitmore J (2002) Coaching for performance. London: Nicholas Brealey Publishing
12. Stowell SJ (1988) Coaching: a commitment to leadership. Train Develop J 42(6):34–39
13. Miller L, Homan M (2002) Ace coaching alliances. Train Develop Am Soc Train Develop 56(1):40–46
14. ICF (2020) ICF. International Coach Federation. https://www.coachfederation.org
15. KCA (2020) KCA. Korea Coach Association. https://www.kcoach.or.kr/
16. Lee S, Kim K, Park H (2016) The teaching and training effect of the treasure talk coaching model on the interaction of early childhood treasures and mothers. J Korean Coach Res 9(2):5–28
17. Fairley SG, Stout CE (2004) Getting started in personal and executive coaching: how to create a thriving coaching practice. Wiley, Hoboken, NJ
18. Boyatzis AR (1982) The competent manager: a model for effective performance. Wiley, New York, pp 20–21
19. Spencer LM, Spencer SM (1993) Competency at work. Wiley
20. Law H (2018) Groups and communities. Handbook of coaching psychology: a guide for practitioners
21. Song YS (2011) A study on outside professional coach's competencies for performance improvement in corporate settings. Korean J Human Res Develop 13(1):53–73
22. Do MH, Kim SY (2019) Development and validation of a coaching competency scale. J Korean Coach Res 12(4):59–81
23. Grönroos C (1984) A service quality model and its marketing implications. Eur J Mark 18(4):36–44
24. Parasuraman A, Zeithaml VA, Berry LL (1988) Servqual: a multiple-item scale for measuring consumer perceptions of service quality. J Retail 64(Spring):12–40
25. Cronin JJ Jr, Taylor SA (1992) Measuring service quality: a reexamination and extention. J Market 56:55–68
26. Shen XX, Tan KC, Xie M (2000) An integrated approach to innovative product development using Kano's model and QFD. Euro J Innov Manage 3(2):91–99
27. Etgar M, Fuchs G (2009) Why and how service quality perceptions impact consumer responses. Manag Serv Qual Int J 19(4):474–485
28. Klaus S (2017) The Fourth Industrial Revolution. World Economic Forum
29. Byun DH (2018) Survey of service industry policy and big data analysis of core technology in preparation of the fourth industrial revolution. J Serv Res Stud 8(1):73–87
30. Passmore J, Fillery-Travis A (2011) A critical review of executive coaching research: a decade of progress and what's to come. Coaching: An International Journal of Theory, Research and Practice 4(2):70–88
31. Bandura A (1986) Social foundations of thought and action: a social cognitive theory. Englewood Cliffs, NJ: Prentice-Hall
32. DiDonato TE, Krueger JI (2010) Interpersonal affirmation and self-authenticity: a test of rogers's self-growth hypothese. Self and Identity 9:322–336
33. Wickham RE (2013) Perceived authenticity in romantic partners. J Exp Soc Psychol 49:878–887
34. PZB (1985) A conceptual model of service quality and Its implication for future research. J Mark 49:41–50

Determinants of Technology-Based Self-Service Acceptance

Seulki Lee and Donghyuk Jo

Abstract The spread of the COVID-19 has amplified the use of non-face-to-face service due to the psychological risk of contact. As a result, non-face-to-face service is being presented in a new paradigm that will lead the post-corona era, and the importance of this will be further increased. In particular, the service industry, which had been face-to-face transaction, has made remarkable progress and is highly utilized in the restaurant industry. Therefore, the purpose of this study is to explore the factors that effects the acceptance intention of technology-based self-service of food and beverage store customers. In the survey, we conducted a questionnaire survey of customers who had used technology-based self-service at coffee shop. Hypothesis test was done using AMOS statistical programs. The results of this study have academic implications for investigating the antecedent variables to the acceptance intention of technology-based self-service in uncertain environments. In practice, it will contribute to decision-making and business strategy formulation for the post-corona era.

Keywords TBSS (Technology-based self-service) · VAM (Value-based adoption Model) · Perceived value · Attitude · Acceptance intention

1 Introduction

COVID-19 has influenced across the world on the way of life including the daily routine of individual. For that, it serves as a motive that the face-to-face service enterprise concentrates on the digitalization for the non-face-to-face service more [1]. Particularly, the digitalization was accelerated in the food service area and retail area, which is because of the fact that COVID-19 spreads by the human contact

---

S. Lee
Department of Business Administration, Sangmyung University, Seoul, Korea
e-mail: sklee4286@gmail.com

D. Jo (✉)
Department of Business Administration, Soongsil University, Seoul, Korea
e-mail: joe@ssu.ac.kr

© The Author(s), under exclusive license to Springer Nature Switzerland AG 2021   39
J. Kim and R. Lee (eds.), *Data Science and Digital Transformation in the Fourth Industrial Revolution*, Studies in Computational Intelligence 929,
https://doi.org/10.1007/978-3-030-64769-8_4

and the social distancing campaign [2], for which the non-face-to-face consumption culture is being spreading. Example of non-face-to-face consumption us self-service encounter, online purchasing, online payment, kiosk, etc. [3].

The food service businesses and the retail businesses, which have invested proactively in the non-face-to-face in our country, showed the pattern that the sales are rather increased during the period that COVID-19 was spreading rapidly. Ono-face-to-face order and payment system of Starbucks Korea can be the representative example. Starbucks Korea is providing 'Siren Order', which is the non-face-to-face service based on the technology that it has developed independently. 'Siren Oder' is the service that the customer can receive the beverage personally after selecting and paying the desired beverage through dedicated APP and can minimize the contact with the employee. The number of order received through 'Siren Order' from January to February, 2020 was more than 8 million and increased by 25% compared to the same period of 2019 [4].

'Siren Order' of Starbucks Korea is a sort of TBSS (Technology-based Self-service) and corresponded to Smart Ordering System. TBSS means the technology service that the consumer can participate directly to the service production through machine or internet, etc. without help of employee [5]. Recently, in the food and beverage store, Smart Ordering System is utilized actively [6]. Smart Ordering System has advantage that it takes short time from ordering to payment through the mobile devices possessed by the individual and can minimize the communication error. However, in the position of business, since the initial cost is incurred by the customized software development by enterprise and authentication [6] and there is no installation-related cost transferred to consumer, if the user wants, the user can stop using the service any time.

Therefore, to TBSS like smart ordering system, forecasting the intention if the user would use relevant technology is essential. In addition, various researches on the kiosk that minimizes the contact through the device installed in specific location have been made but the research on the smart ordering system, which TBSS is realized through mobile devices carried by user is not enough.

Therefore, the purpose of this study is to verify the factors having impact on the TBSS acceptance intention empirically focusing on the smart ordering system. To do that, this study established the research model based on the value-based adoption model saying that considering the technology user as consumer, the value perceived from technology would increase the technology acceptance intention, and to examine the relations by drawing key factors based o preceding research. In the meantime, since most of the smart ordering systems in the food and beverage store do not incur the initial installation cost in the user, have been introduced for the convenience of use and its technology is not complicate, the key factor was limited to the benefits.

The results of this study are expected to provide the information what are the factors that the users consider as important to the organization, which wants to introduce the smart ordering system as a part of TBSS at the time of preparing post COVID era.

2 Theoretical Background and Hypotheses

2.1 TBSS (Technology-Based Self-Service)

TBSS means the service that allows the consumer to participate directly in the service production through the machine or technology such internet, etc. without help of employee [5]. Touch screen, self-scanning, ATM, internet banking, mobile banking, chatbot, etc. are considered as TBSS. Recently, as the mobile devices like smart phone become importance interface of TBSS by the increase of availability of Wi-Fi, Bluetooth, etc., new mobile-based TBSS is emerged [7]. Therefore, in the food service industry, the mobile-based smart ordering system is included as a type of TBSS [6].

Smart ordering system composed of 3 elements. First, the user must have Web Ordering System in his/her smart phone. Next, it should have Menu Management System to be able to select menu and Order Retrieval System that the employee can see the order details ordered by user [8]. As such, the smart ordering system has a hassle that the user must install the system in the personal mobile device.

Nevertheless, the reason why the user prefers to TBSS like smart ordering system is because of the speed and convenience taken to make order, cost saving, fun, minimization of communication error, minimization of unnecessary meeting with employee, etc. [5, 6]. In our country, "Siren Order' of Starbucks can be the representative example of smart ordering system. 'Siren Order', which was TBSS developed independently by domestic Starbucks, is the service that the customer can select and receive the desired beverage after payment through dedicated APP. Particularly, since it is designed to select the personal option for product, the individual customized service is possible. In addition, through the APP notice, the information such as various event benefits, promotional beverage, etc. are provided [9]. Therefore, in this study, the factors having influence on the mobile device-oriented TBSS acceptance intention will be examined using Siren Order of Starbucks.

2.2 VAM (Value-Based Adoption Model)

As the representative theory on the process that the user is accepting specific technology, there is TAM (Technology Acceptance Model) suggested by Davis et al. [10]. However, TAM has limitation that it could not include the general consumer as a subject of accepting technology. To overcome such limitation, VAM (Value-based Adoption Model) was emerged trying to explain by what the service consumer accepts the technology [11]. VAM defined the IT technology user as consumer and focused on maximizing the personal value of the consumer [11]. Kim et al. [11] tried to explain the acceptance of technology and service based on the concept of perceived value by Zeithaml [12]. Therefore, the perceived value from the new technology becomes importance antecedent for technology acceptance intention.

In the meantime, the consumer experiences the benefit and sacrifice obtained by using IT technology before perceiving comprehensive value. Therefore, both the benefit and sacrifice should be considered as an element having influence on the value in accepting the technology and in VAM, the usefulness and enjoyment were suggested as element of benefit and the technicality and perceived fee were as the element of sacrifice [11, 13].

The usefulness means the degree that new technology is useful in the daily life or business, etc. and it is the similar concept to the functional advantage of product or service use [11, 14]. The usefulness of the technology can improve the ability of carrying out the work and makes the decision making easy [14]. Therefore, the more the new technology is perceived as useful, the higher the value of the technology can be perceived, which is proved in the research related to the multiple technology acceptance [11, 14, 15].

Enjoyment means the degree of pleasure, joy and satisfaction felt using new technology [10, 11]. The person who experiences the pleasure and joy by using the technology is highly likely to accept the technology and use widely by perceiving the value of the technology highly [10]. Therefore, the multiple researches utilize the enjoyment as major predictor of the perceived value [15–17].

Technicality is the degree of perceiving technically superior in the process of providing services and is determined by the system reliability, connectivity and efficiency [11]. If in the position of technology user, the use of technology is felt difficult and the time to connect it is taken longer, it is recognized as cost and causes the negative mental state. However, on the contrary, if the use of technology is not complicate and easy, the positive value can be formed. The research results that the complexity of the technology has negative influence on the perceived value supports it [15].

Perceived cost is paying the monetary consideration, which includes the actual price of the product or service, and measured with the consumer's perception on the cost actually paid [11]. If the cost paid to use the technology is perceived high, negative perception can be formed. On the contrary, the more the cost is saved or additional benefit is increased, the perceived value can be increased [18]. As the smart ordering system, the object of this study, does not generate the cost related to use of system besides the existing price of product ad provides the benefit such as accumulation of point, issuance of coupon, etc., it will be used by changing to cost advantage. In the meantime, the research by Kwon and Seo [19] showed that the economic feasibility has positive influence on the perceived value.

Through the aforementioned contents, it was expected that the usefulness, enjoyment, technicality and the cost advantage would have influence on the perceived value of the technology-based self-service in the food and beverage store and following hypothesis was drawn.

H1. Perceived benefit will have a positive effect on the perceived value.

Perceived value means overall evaluation on the product or service utility [12] and is emphasized as a key antecedent determining the consumer's attitude or behavior [20]. For example, the research that explains the acceptance of the service combined with new technology such as mobile hotel reservation system [21], IPTV [22], e-book subscription service [23], IoT smart home service [24], VR [15], etc. proved that the perceived value is important predictor of acceptance intention. Therefore, it can be expected that in case of the user perceiving the value of TBSS, the acceptance intention would be increased and following hypothesis was established.

H4. Perceived value will have a positive effect on acceptance intention.

2.3   VAB (Value-Attitude-Behavior) Model

As the model predicting consumption behavior of the consumer, there is VAB (Value-Attitude-Behavior) Model. VAB Model suggested by Homer and Kahl [25] is the model useful in predicting the consumption behavior and is the theory that the consumer's behavior is formed through the value and attitude.

Attitude in VAB Model plays the mediating role between the abstract value and specific behavior [25]. Attitude is the learned emotional inclination relatively formed consistently on the specific target and means the psychological reaction appeared favorably or unfavorably through the experience [26]. Dabholka and Bagozzi [27] defined the attitude as the experiential evaluation such as good or bad, pleasant or unpleasant, etc. Therefore, in order to form the attitude, the cognitive reaction like value perception should be preceded.

Since the value means overall evaluation on the utility of product or service according to Zeithaml [12], it is viewed as single dimension but is viewed as multi-dimensional perspective depending on the researcher [28]. While the approach to the single dimension has overall character of practical aspect, in case of approaching to multi-dimension, it has strong complex characters of psychological, cognitive and emotional dimension [29]. In the multi-dimensional structure, utilitarian value, hedonic value, acquisition value, transaction value, use value, functional value, emotional value, social value, etc. are used [28, 30, 31]. If multi-dimensional structure is examined, it is observed that the aspect of diverse benefits obtained by the experience is considered.

Therefore, it can be expected that the benefit element of technology use drawn according to VAM (usefulness, enjoyment, technicality, cost advantage) also have significant influence on the attitude. Gan and Wang [32] examined the influence on the attitude by establishing the benefits of social commerce as value. Therefore, following hypotheses were drawn.

H2. Perceived benefit will have a positive effect on the attitude.

H3. Perceived value will have a positive effect on the attitude.

The attitude influenced by the value influences on the consumer's behavior [24]. Dabholka and Bagozzi [27] verified that the behavior life decision making is influenced by the attitude. Karjaluoto et al. [33] suggested in the research on the mobile banking acceptance intention that the attitude can become a powerful antecedent, and Alotaibi [34] proved that the attitude has influence on the cloud computing acceptance. The research on Home Smart IoT service by Kim et al. [24] demonstrated that the attitude is the major influential factor of the acceptance intention, through

Table 1 Sample characteristics

Category and items		Sample size	Ratio (%)
Gender	Female	127	53.8
	Male	109	46.2
Age	20–29	74	31.4
	30–39	109	46.2
	40–49	40	16.9
	More than 50	13	5.5
Frequency of use (recent 6 month)	1–2	100	42.4
	3–4	46	19.5
	5–6	26	11.0
	More than 7	64	27.1

which it was inferred that the attitude would improve the intention of accepting new technology and following hypothesis was drawn.

H5. Attitude will have a positive effect on the acceptance intention.

3 Research Method

3.1 Sample and Data Collection

In order to verify the proposed model, we conducted a questionnaire survey for customers who have used the smart order system (siren order) in Starbucks Korea. As a result of the survey, 250 questionnaires were collected, and 236 cases were selected as valid samples after eliminating missing or inadequate data. The samples of this study are summarized in Table 1.

3.2 Measures

For the content validity of the measurement tool, measurement items that have already been verified in previous studies were derived and modified to suit the purpose of this study. In this study, usefulness, enjoyment, technology, cost advantage, perceived value, attitude, and acceptance intention were set as the main concepts. For tools to measure the concept, studies by Kim et al. [11], Lin et al. [22], Yu et al. [14], and Kim et al. [24] were referenced, and then measured them by the Likert Five-point scale (from Not at all to Very Much). The measurement items in this study are summarized in Table 2.

Table 2 Confirmatory factor analysis based on reliability

Variable	Measurement items	Factor L.D.	C.R.	Crb. Alpha
Usefulness	Reduction of time required	0.783	0.944	0.893
	Helped to achieve the desired purpose	0.845		
	Usefulness of the order method	0.836		
	Usefulness of payment method	0.836		
Enjoyment	Amusement	0.882	0.897	0.895
	Attractive	0.824		
	Pleasure	0.876		
Technicality	Ease of use	0.798	0.914	0.868
	Available anywhere	0.861		
	Less occurrence of problems	0.829		
Cost advantage	Economic benefits	0.793	0.898	0.811
	Reasonable price	0.864		
Perceived value	Value for money invested	0.828	0.921	0.879
	Benefits for the effort invested	0.837		
	Overall value	0.856		
Attitude	Prefer ordering through Siren order	0.643	0.901	0.770
	Satisfaction with the services provided	0.787		
	Overall positive experience	0.767		
Acceptance intention	Want to use it in the future	0.836	0.903	0.826
	Will be used when ordering in the future	0.793		
	Want to use even if there are competitors	0.722		

3.3 Analysis Method

For hypothesis testing, we first carried out measurement model analysis for validity and reliability analysis, and then structural equation model analysis using Amos v.22.0 program.

Table 3 Discriminant validity

Variable	1	2	3	4	5	6	7
1. Usefulness	**0.899ᵃ**						
2. Enjoyment	0.115	**0.862ᵃ**					
3. Technicality	0.093	0.196	**0.884ᵃ**				
4. Cost advantage	0.491	0.134	0.166	**0.903ᵃ**			
5. Perceived value	0.393	0.217	0.564	0.489	**0.892ᵃ**		
6. Attitude	0.490	0.150	0.411	0.508	0.596	**0.868ᵃ**	
7. Acceptance intention	0.406	0.176	0.486	0.493	0.649	0.579	**0.871ᵃ**

Note ᵃsquare root of AVE (Average Variance Extract)

4   Analysis and Results

4.1   Measurement Model

Confirmative factor analysis was conducted to secure the reliability and validity of the measurement tool. For this, standard $\chi^2(\chi^2/df)$, RMSEA, GFI, NFI, TLI, CFI were used to check goodness of fit. As a result of confirmatory factor analysis of measurement model, $\chi^2/df = 1.21(\leq 3)$, RMSEA $= 0.03(\leq 0.05)$, GFI $= 0.927(\geq 0.9)$, NFI $= 0.934(\geq 0.9)$, TLI $= 0.985(\geq 0.9)$, CFI $= 0.988(\geq 0.9)$ all indicators were found to be suitable. After verifying measurement model's fitness, reliability and validity were analyzed. Reliability is examined by accessing construct reliability (C.R.). C.R. ranges from 0.897 to 0.944, exceeding the cut-off value of 0.7 for C.R. Validity is was evaluated by average variance extracted (AVE), and it was found to be 0.744–0.815, exceeding the standard value of 0.5. Additionally, to identify the discriminant validity, the square root of the AVE for each construct is compared with the correlation coefficients between two constructs [35, 36]. When the AVE square root is larger than the correlation coefficient, it is determined that the discrimination validity is secured. As a result of analysis, reliability and validity were verified and the detailed results are presented in Tables 2 and 3.

4.2   Structural Model

The fitness of the measurement model and the reliability and validity of the measurement tool were confirmed to be normal, and the structural model was analyzed. As a result of structural model's fitness test, $\chi^2/df = 1.225(\leq 3)$, RMSEA $= 0.031(\leq 0.05)$, GFI $= 0.925(\geq 0.9)$, NFI $= 0.932(\geq 0.9)$, TLI $= 0.984(\geq 0.9)$, CFI $= 0.987(\geq 0.9)$, all of which were above the baseline.

4.3   Hypotheses Tests

Structural equation model analysis was performed to verify the hypothesis of the proposed model. As a result, first, Usefulness ($\beta = 0.149$, C.R. $= 2.210$), Technicality ($\beta = 0.548$, C.R. $= 8.572$) and Cost advantage ($\beta = 0.386$, C.R. $= 5.518$) have a positive effect on Perceived value, therefore, H1a, H1c and H1d were supported. On the other hand, Enjoyment ($\beta = 0.047$, C.R. $= 0.887$) did not have a positive effect on Perceived value, therefore, H1b was not supported. Second, Usefulness ($\beta = 0.265$, C.R. $= 3.425$), Technicality ($\beta = 0.246$, C.R. $= 2.631$) and Cost advantage ($\beta = 0.278$, C.R. $= 2.903$) have a positive effect on Attitude, therefore, H2a, H2c and H2d were supported. However, Enjoyment ($\beta = -0.018$, C.R. $= -0.315$) did not significantly affect Attitude, therefore, H2b was not supported. Third, perceived value($\beta = 0.289$, C.R. $= 2.422$) was shown to have a positive effect on attitude, and Hypothesis 3 was supported. Fourth, perceived value($\beta = 0.482$, C.R. $= 4.875$) was shown to have a positive effect on acceptance intention, therefore H4 was supported. Finally, it was confirmed that attitude($\beta = 0.401$, C.R. $= 3.864$) had a significant positive(+) effect on acceptance intention. Thus, H5 was supported. The results of hypotheses test are summarized in Table 4.

Table 4   Results of hypotheses tests

H	Path	Estimate (β)	C.R. (t)	Result
H1a	Usefulness → Perceived value	0.149	2.210*	Supported
H1b	Enjoyment → Perceived value	0.047	0.887	Not supported
H1c	Technicality → Perceived value	0.548	8.572***	Supported
H1d	Cost advantage → Perceived value	0.386	5.518***	Supported
H2a	Usefulness → Attitude	0.265	3.425***	Supported
H2b	Enjoyment → Attitude	−0.018	−0.315	Not supported
H2c	Technicality → Attitude	0.246	2.631**	Supported
H2d	Cost advantage → Attitude	0.278	2.903**	Supported
H3	Perceived value → Attitude	0.289	2.422*	Supported
H4	Perceived value → Acceptance intention	0.482	4.875***	Supported
H5	Attitude → Acceptance intention	0.401	3.864***	Supported

$*p < 0.05$; $**p < 0.01$; $***p < 0.001$

5 Conclusions

5.1 Summary and Discussion of Results

This study intended to draw the influential factor based on the VAM (Value-based Adoption Model) and VAB (Value-Attitude-Behavior) model and to identify the structural relations among variable in order to identify the factors of promoting the user's acceptance intention of TBSS, which is receiving the attention by the risk recognition on the contact due to COVID 19 virus. Particularly, Smart Ordering System, which is being introduced actively in the store specialized in beverage of food service sector was examined intensively. Therefore, the study was performed focusing on 'Siren Order' of Starbucks Korea, which was developed independently by the enterprise and already has multiple users. The summary of research results are as follows.

However, the enjoyment appeared not to have significant influence on the perceived value and the attitude, which means the more it is believed that Smart Ordering System would help the user in the life by using it, the easier how to use and the less failure and the more it is perceived that the economic benefits are provided, the more the value perceived comprehensibly and the attitude toward the technology are improved. Second, it was confirmed that the perceived value has significant positive (+) effect on the attitude, which means that the better the value of new technology, the more positive attitude is formed. Third, as the perceived value and the attitude appeared to have significant positive (+) effect on the acceptance intention, it supports the results of preceding researches [15, 21–24, 33, 34].

5.2 Implications and Limitations

The academic implications of this study are as follows. First, in the circumstance that the empirical research on Smart Ordering System, a kind of TBSS, is not enough, the theoretical system, that explains the intention that the user would accept the relevant technology, was prepared. Second, this study has meaning that the influential relations among the concepts were identified by applying VAM (Value-based Adoption Model), which examines the influential factor in the aspect of perceived value by the user in accepting new technology, and VAB (Value-Attitude-Behavior) Model, which the consumer's behavior is receiving the influence by the value and the attitude.

Practical implications are as follows. First, from the fact that the usefulness, technicality, cost advantage suggested as benefit by the technology in VAM have positive influence on the perceived value and the attitude toward Smart Ordering System, the need of design to be able to emphasize the usefulness, technicality, cost advantage is suggested. For example, diverse methods are introduced in order to satisfy the individual need in the ordering method and payment method through Smart Ordering System and at the same time, the operating method should be designed conveniently. In addition, considering that the cost advantage has significant influence on the value perception and attitude formation, it is necessary to perform the service or promotion that can provide the monetary benefits only from relevant system. In case of "Siren Order' of Starbucks Korea, it provides the benefit in the aspect of cost to user through the issuance of BOGO coupon $(1 + 1)$, the free beverage through the accumulation of Star, performance of specific event, etc. Second, since the enjoyment, which was proved in multiple researches, represented not to have significant influence on the perceived value and the attitude, it can be estimated that it would not be the major factor in the smart ordering system, of which purpose of use is simple and clear. Therefore, if the food service sector would introduce smart ordering system, it is necessary to establish the strategy considering the aspect of usefulness, technicality and cost advantage with priority than the elements related to the enjoyment. Third, it was confirmed that the more the perceived value and the attitude toward the technology are positive, the higher the user's intention to accept the smart ordering system out of TBSS, through which the need to seek out the measures to induce the acceptance can be suggest by stimulating the aspect of user's value and attitude.

The limitations of this study and the direction of follow-up research are as follows. First, to understand the non-face-to-face consumption culture at the point preparing post-Covid-19 era, the aspects of benefit and value received from new technology were examined from the smart ordering system, which is a kind of TBSS. However, the innovation of individual makes to take more active action in accepting new technology [14]. Therefore, it is regret that in this study, the innovative disposition possessed by the individual user was not considered, and in the follow-up research, the characteristics of the individual before contacting with new technology should be considered. Second, since this study was performed focusing on 'Siren Order' of Starbucks Korea, which is the Smart Ordering System introduced for the first time in the domestic food and beverage store, it is necessary to make effort to generalize the research results by performing the comparison with the system introduced by other brand or the integrated research.

References

1. Diebner R, Silliman E, Ungerman K, Vancauwenberghe M (2020) Adapting customer experience in the time of coronavirus. McKinsey & Company. Accessed 19 May 2020
2. Shahbaz M, Bilal M, Moiz A, Zubair S, Iqbal HM (2020) Food safety and COVID-19: precautionary measures to limit the spread of coronavirus at food service and retail sector. J Pure Appl Microbiol 14(Special Edition):1–9
3. Lee SM, Lee D (2020) "Untact": a new customer service strategy in the digital age. Serv Bus 14(1):1–22
4. Starbucks Korea. http://starbucks.co.kr/bbs/getBodoView.do?seq=3767 (2020)
5. Meuter ML, Ostrom AL, Roundtree RI, Bitner MJ (2020) Self-service technologies: understanding customer satisfaction with technology-based service encounters. J Market 64(3):50–64
6. Umap S, Surode S, Kshirsagar P, Binekar M, Nagpal N (2018) Smart menu ordering system in restaurant. IJSRST 4(7):207–212

7. Tanimura M, Ueno T (2013) Smartphone user interface. Fujitsu Sci Tech J 49(2):227–230

8. Mishra N, Goyal D, Sharma AD (2018) Automation in restaurants: ordering to robots in restaurant via smart ordering system. Int J Technol Manage 4(1):1–4

9. Baek SH, Yim HR, Kim HS (2019) A study of the acceptance intention of customers in coffee franchise using extended technology acceptance model (E-TAM): focused on starbucks mobile application. Culinary Sci Hospitality Res 25(4):146–155

10. Davis FD, Bagozzi RP, Warshaw PR (1989) User acceptance of computer technology: a comparison of two theoretical models. Manage Sci 35(8):982–1003

11. Kim HW, Chan HC, Gupta S (2007) Value-based adoption of mobile internet: an empirical investigation. Decis Support Syst 43(1):111–126

12. Zeithaml VA (1988) Consumer perceptions of price, quality, and value: a means-end model and synthesis of evidence. J Market 52(3):2–22

13. Chunxiang L (2014) Study on mobile commerce customer based on value adoption. JApSc 14(9):901–909

14. Yu J, Lee H, Ha I, Zo H (2017) User acceptance of media tablets: an empirical examination of perceived value. Telematics Inform 34(4):206–223

15. Vishwakarma P, Mukherjee S, Datta B (2020) Travelers' intention to adopt virtual reality: a consumer value perspective. J Destination Mark Manage 17:1–13

16. Yang H, Yu J, Zo H, Choi M (2016) User acceptance of wearable devices: an extended perspective of perceived value. Telematics Inform 33(2):256–269

17. Kim SH, Bae JH, Jeon HM (2019) Continuous intention on accommodation apps: integrated value-based adoption and expectation-confirmation model analysis. Sustainability 11(6):1578

18. Lovelock CH (2001) Services marketing. Prentice Hall, Upper Saddle River, NJ

19. Kwon HK, Seo KK (2013) Application of value-based adoption model to analyze SaaS adoption behavior in Korean B2B cloud market. Int J Advancements Comput Technol 5(12):368–373

20. Tam JL (2004) Customer satisfaction, service quality and perceived value: an integrative model. J Mark Manage 20(7–8):897–917

21. Wang HY, Wang SH (2010) Predicting mobile hotel reservation adoption: insight from a perceived value standpoint. Int J Hospitality Manage 29(4):598–608

22. Lin TC, Wu S, Hsu JSC, Chou YC (2012) The integration of value-based adoption and expectation–confirmation models: an example of IPTV continuance intention. Decis Support Syst 54(1):63–75

23. Hsiao KL, Chen CC (2017) Value-based adoption of e-book subscription services: the roles of environmental concerns and reading habits. Telematics Inform 34(5):434–448

24. Kim Y, Park Y, Choi J (2017) A study on the adoption of IoT smart home service: using value-based adoption model. Total Qual Manage Bus Excellence 28(9–10):1149–1165

25. Homer PM, Kahle LR (1988) A structural equation test of the value-attitude-behavior hierarchy. J Pers Soc Psychol 54(4):638–646

26. Aaker JL (1997) Dimensions of brand personality. J Mark Res 34(3):347–356

27. Dabholkar PA, Bagozzi RP (2002) An attitudinal model of technology-based self-service: moderating effects of consumer traits and situational factors. J Acad Mark Sci 30(3):184–201

28. Sweeney JC, Soutar GN (2001) Consumer perceived value: the development of a multiple item scale. J Retail 77(2):203–220

29. Sanchez J, Callarisa L, Rodriguez RM, Moliner MA (2006) Perceived value of the purchase of a tourism product. Tour Manag 27(3):394–409

30. Peña AIP, Jamilena DMF, Molina MÁR (2012) The perceived value of the rural tourism stay and its effect on rural tourist behaviour. J Sustain Tour 20(8):1045–1065

31. Shan G, Yee CL, Ji G (2020) Effects of attitude, subjective norm, perceived behavioral control, customer value and accessibility on intention to visit Haizhou Gulf in China. J Mark Adv Pract 2(1):26–37

32. Gan C, Wang W (2017) The influence of perceived value on purchase intention in social commerce context. Internet Res 27(4):772–785

33. Karjaluoto H, Püschel J, Mazzon JA, Hernandez JMC (2010) Mobile banking: proposition of an integrated adoption intention framework. Int J Bank Market 28(5):389–409

34. Alotaibi MB (2014) Exploring users' attitudes and intentions toward the adoption of cloud computing in Saudi Arabia: an empirical investigation. J Comput Sci 10(11):2315–2329
35. Fornell C, Larcker DF (1981) Evaluating structural equation models with unobservable variables and measurement error. J Mark Res 18(1):39–50
36. Ping RA Jr (2004) On assuring valid measures for theoretical models using survey data. J Bus Res 57(2):125–141

Pre-verification of Data in Electronic Trade Blockchain Platform

Saeyong Oh, Sanghyun Cho, Sunghwa Han, and Gwangyong Gim

Abstract Blockchain technology has several advantages, including security, transparency, and stability. It can reduce data management cost and be applied to important electronic document related services. However, in the perspective of the subject, in order to apply the blockchain technology to electronic document-related services form, it is necessary to review the integrity of the electronic documents to verify that a stable transaction is possible. In addition, in the perspective of the consumer, because electronic documents contain content related to their important personal information or contracts, electronic documents should be proved to be stable in the blockchain trading platform. Therefore, this study aims to present an architecture that can verify the integrity of the blockchain-based electronic document platform by presenting a pre-validation method for electronic documents. Furthermore, by establishing an actual pilot system based on this architecture, a system where customers can reliably understand their transactions and contracts was made.

Keywords Blockchain · Electronic trade platform · Data pre-verification · Data efficiency · Data integrity · File format verification · Data forgery · Electronic document

S. Oh · S. Cho · S. Han · G. Gim (✉)
Department of IT Policy and Management, Soongsil University, Seoul, South Korea
e-mail: gygim@ssu.ac.kr

S. Oh
e-mail: ohsy@satu.co.kr

S. Cho
e-mail: cshyms74@naver.com

S. Han
e-mail: taifanz@naver.com

1   Introduction

The scalability of the blockchain technology can be seen as a technology that can solve the problem of efficiency and reliability of the storage and management of electronic document data such as various records, medical data, and contract documents in daily life [1–3].

Electronic documents are typically stored and managed by the entity providing the services, and this centralized structure is a significant burden to information management entities such as businesses and public institutions. This is because the volume of electronic documents increases in the contemporary society and the cost of solving security problems increases. Distributed network systems in blockchain technology can be the next-generation core technology to address these costs [3–5]. There are many things to consider when the provider of e-document-related services decide to provide it services based on the blockchain technology. The primary provider should validate the existing electronic document data. Blockchain cannot be deleted or modified after upload, so the information subject must be able to determine whether electronic documents are valid data before the data is transferred. Next, the integrity of electronic documents on the blockchain should be checked to verify that forged data are negotiable in the distribution of electronic documents. Finally, a system that can check transaction details or transaction information should be established so that users can have faith in the blockchain technology.

This study attempts to experiment with the above three factors and suggest solutions. First, for verification of validity of electronic documents, we would like to select the data format that accounts for the majority of electronic documents. Also, we will suggest a method to verify format verification through the format analyzer of electronic documents, and to verify the damage of files using the data validation model. Second, blockchain is said to be theoretically impossible to falsify data and safe from hacking threats. However, there has been no research that has actually verified it with electronic document files. Therefore, it is necessary to verify various variables of electronic document forgery and verify them in the blockchain. To do so, we would like to check the integrity of the data by selecting a blockchain that can be applied to electronic transactions and verifying it experimentally. Third, because the blockchain is in code form, it is difficult for ordinary users to use it, so we would like to create a pilot system that looks like a service portal or transaction system by visualizing transaction details or contracts [4, 6, 7].

2   Theoretical Background

2.1   Advent and Features of Blockchain

A blockchain can be defined as a block configuration that holds records of data in the form of a bundle of individual transactions in a distributed transaction. The block maintains a growing list of data records protected from tampering and forgery. Each of these blocks includes a timestamp and a link to the previous block [8, 9, 10].

In short, the blockchain is a network-based distributed system [11, 12]. The most distinctive feature of blockchain is that it is a distributed book-operating structure without a central management entity acting as a broker for the transactions. It has a structure that directly transmits and receives data and shares related information between each node participating in the network without a central control system. This structure eliminates centralization of data, maintains security through consensus algorithms such as strong encryption and proof of work, and distributes records across all nodes in the network, enabling transparency in transactions and stable operation of the system [4, 13, 14].

2.2 Structure of Blockchain

Blockchain can literally be described with blocks and chains. Each block records the data exchanged between all network participants and each other, and these blocks are linked by a chain. Most transaction-related information is contained in the Transaction Details section. Any change in transaction information will result in different hash values [15]. Changes can be easily detected if a block of data is arbitrarily modulated. The block contains the hash number of the previous block as well as its own hash value. Because of this connection, blocks are linked together to form chains, and the longer the chain is, the more difficult it is to forge [16–18].

2.3 Advanced Research

Blockchain technology has the scalability to be utilized in a variety of areas [4, 19, 20]. This section provides an overview of the active research using blockchain technology in diverse fields. First, Civelek and Ozalp [21] stated that research on blockchain technology and paperless trade proves that the blockchain technology can provide document integrity in foreign trade and finance. The study argued that changes such as existing legal problems and infrastructure should be integrated into the platform.

Ghazali and Saleh [22] designed a model that verifies the certificate by requesting it to a public infrastructure-based blockchain system by generating hash values using hash algorithms and generating private keys and timestamps in order to give it an essential element of the blockchain. Through this, the author suggested the possibility that blockchain technology can be used for degree certification.

Nizamuddin et al. [23] proposed a smart contract code solution to avoid security hacking by developing a solution to store documents in distributed systems and control versions of documents in distributed document version control studies using Ethereum blockchain and IPFS (InterPlanetary File System).

3 Empirical Analysis

3.1 Efficiency of Pre-verification Model for Electronic Document Data

In the stage of verifying the validity, two stages of experiment were done to verify the file format of electronic document and the validity of the file itself. The experiment used language C/C++, and in the text verification stage, because there are many text format types, it is hard to verify all the documentation formats. But the experiment tried to embody test model that can verify a broader range of formats. Even if a document format is distinguished and verified as a document that can be opened properly at the stage of verifying the validity of documents, because there is no method to open the damaged bit stream or encrypted electronic documents, this part was exempted in the experiment.

Document File Format Verification

The e-document format analyzer built in this study was targeted at the following Hancom format, MS office-related format, and image format, and was conducted by extracting identifiable information from the headers of the text and attachments in the document to verify the format and comparing it with the DB containing the format information.

As seen in Table 1, each file format was normally recognized.

Document's Efficiency Verification

The file's efficiency verification tool developed in relevant study is designed to allow the file format to perform an effectivity verification only on the certified format, and the efficiency verification is possible if the added file format can be verified in the file format analyzer. The tool verifying file's efficiency is configured to check if the file is damaged, and the file's damage types are mainly as follows.

- File of which the name exists but content is empty.
- Documents that cannot be opened due to partial damage.
- Documents that cannot be opened because there is a password on the file.
- Document that cannot open file because file header is damaged.

As seen above there were some files that were damaged. These files were judged to not be worth keeping it because there is no way to open it, and this is the reason

Division	File format(extension)	Test result
Hangeul	hwp(3.0, 4.0, 5.0)	Normal
MS OFFICE	xlsx, xls, docx, doc, pptx ppt	Normal
Image	png, jpeg, gif	Normal
PDF	PDF	Normal
Attached file	hwp, doc, pdf	Normal

Table 1 The result of document format verification experiment

Fig. 1 Efficiency
verification algorithm of
electronic document data

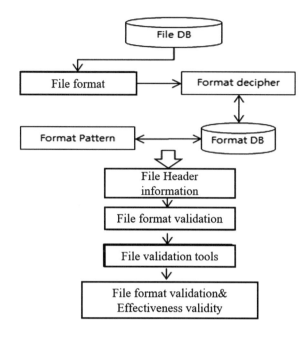

why file's efficiency verification is needed. For the design of the file's efficiency verification tool, the verification tool was designed based on the architecture as shown in Fig. 1, and its configuration is as follows.

The experiment was conducted on the electronic document validation tool, and the sampled text document selected 50 files from the document format analyzer, of which 20 files were selected, each with an empty file, a file with a damaged document, a file with a password, and a file with a damaged document header. The remaining files were separated into normal documents to validate the electronic document.

As Table 2 shows, the test confirmed that all 50 files were normally validated. The empty files and password-set files, and the document header showed accurate classification results for five corrupted files and parts of the files, and all the other normal files were classified as normal.

3.2 The Result of Integrity Verification on Electronic Document Data

To verify the integrity of the blockchain of electronic documents, the appropriate blockchain source for electronic document transactions was selected and the test environment was composed of test servers and nodes based on the selected source.

Table 2 Document's efficiency verification

Division	File name	Extension	File size (KB)	Whether test data error exists	Verification result	Verification contents
1	Hangeul1	hwp	15	Normal	Success	Normalfile
2	Hangeul2	hwp	20	Normal	Success	Normalfile
3	Hangeul3	hwp	15	Error	Empty document	Empty document detection
4	PDF1	pdf	30	Normal	Success	Normalfile
5	PDF2	pdf	30	PASSWORDS	passwords	Passwords document detection
6	PDF3	pdf	50	Normal	Success	normalfile
7	XLS1	xls	35	Normal	Success	normalfile
8	XLS2	xls	40	Error	File damage	Damaged file detection
9	XLS3	xls	44	Normal	Success	Normalfile
10	DOC1	doc	42	Error	Empty document	Empty document detection
:						
50	PPTX3	pptx	85	Normal	Success	Normalfile

Creating blockchain

First, a blockchain is created. On the first server called chain1, parameter values of the blockchain are generated in the form of data files and have a set value to execute the blockchain in the params.dat file. Afterwards, the chain is operated after checking the initial setting of the params.dat file, and once the blockchain is created on the node, it will have basic specifications to prepare network activities such as radio waves and connections. Related information was organized in Table 3.

Connecting blockchain

Next, based on the same settings on the second server after the creation of the blockchain, the blockchain node can be connected to the first chain1. The first node created for authorization after initialization is completed, the Wallet Address is

Table 3 Structure of blockchain creation

Basic information			
Name	Chain1	Version	1.0
Protocol	10,008	Node address	chain1@xxx.xxx.xxx.:xxxx
Permissions (authority)	Connection, Transmit, receive, etc.	Wallet address	1NnPLDUBauNxqB..38digits Numeric, English combination code

Fig. 2 Structure of blockchain connection

approved, and each node in the blockchain is approved by P2P (Peer to Peer) in the same way as Fig. 2 between each node, rather than by the server or client.

These configured nodes become peers and form networks. Also, the number of peers can continue to increase.

Stream Test

Stream is a very important function of the blockchain-based on multi-chain system currently tested. It is a functional area for documents and searches that are used as smart contracts for Ethereum, not as asset transactions, where data such as documents and values that exist inside the blockchain are stored here.

Counterfeit verification test containing electronic document

Electronic document verification tests were conducted by registering documents directly through web-implemented demonstrations and process components designed with original reference data. First, the model designed to use the blockchain system in the fake verification test of electronic documents was tested under the assumption that the data generated based on the following three items were the original: the data to which the original is based shall be defined as the time stamp of the document, the META information of the document, and the textual information of the body. The process to construct the verification model is designed as shown in Fig. 3. The registration process can be targeted at any of the basic archives, or any newly generated documents, and the registration process is based on the data generated from the original base data items described above. Electronic Document A is encrypted through the HASH-256 algorithm, including timestamp, document META information and body information of the above documents, converted to the final KEY value to be registered and stored in the blockchain network.

The verification process is shown in Fig. 4 and when the user requests verification of Document A-1, which requires comparison and verification, the blockchain creates the document file information of the document (A-1) for priority comparison through hash value A-1. Then it performs key verification through KEY1 stored in the stream (STREAM) inside the blockchain. After that, forgery is confirmed in case of discrepancy, and the blockchain cannot register new past documents because it cannot delete or modify past block data that is signed inside.

In order to conduct verification tests based on electronic documents, we have clearly identified the forged electronic documents in both the case of forging META information in documents and the case of forging text in the body of documents. In this process, multi-chain APIs were used.

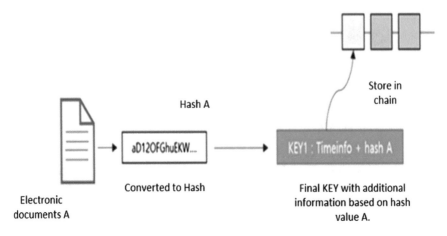

Fig. 3 Registration process of original data

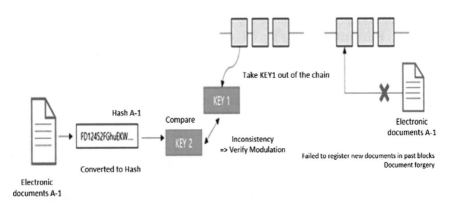

Fig. 4 Process of data verification

Figure 5 also shows the forged META information in the same document. The first document has META information set to USER, the following documents have META information set to OTHER, and the text in the body is the same.

Leverage the previously generated stream to store the information from Document A in the stream. Create a stream and publish Document A to the stream. Stream registration generates transactions in the blockchain and stores data and key values.

After registration, the original data of document A stored is verified by comparing meta information with other document B. Register and certify Document B in the same process.

Through the process, in the event of forgery or tampering with META information in the original information, the information and document B registered in the chain

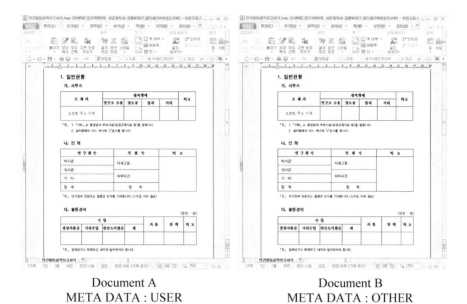

Document A   Document B
META DATA : USER   META DATA : OTHER

Fig. 5 Test sample that checks the modification of META-DATA in the blockchain

are judged to be different data and will not be finally verified. In addition, verification was carried out through modification of the text of the same document, specifically, after setting the META information of the two documents to be verified the same, the contract amount for Document A was 100 as shown in Fig. 6, the contract amount for Document B was calculated to be 500, and part of the document contents was modified. Similarly, the return value was returned to zero and the body text was identified as modified.

The above verification process was able to determine the integrity of data through an electronic document forgery experiment applied with blockchain technology and confirmed that the scenario of blockchain-based electronic document service within the scope of this study can be used for actual services.

Realization of user based blockchain trade system

In this section, as mentioned in the previous chapter, a pilot system that is readily available to the general consumer is embodied. The pilot system has established the verification and smart contract of electronic document files through the blockchain system so that consumers can directly search and verify them. This can be identified through Table 4 and Fig. 7. Table 4 is a detailed configuration of the pilot system, and Fig. 7 shows the front screen on which the pilot system is shown to the consumer.

The first RAW DATA transferred to the blockchain can verify transaction point information through the Time Verification module at each transaction. In addition, the first RAW DATA can verify file modifications through a file validation model. File Viewer is a file converted to SVG to serve consumers, not original data, and provides point-of-trade certification and file validation marks within the viewer as

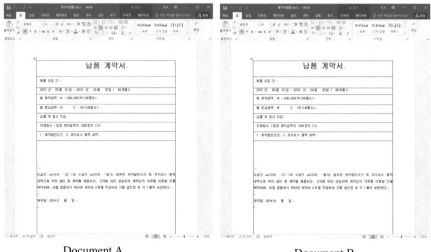

Document A   Document B
Contract amount: 100   Contract amount: 500

Fig. 6 The test sample that applicate modification in text

Table 4 The pilot system that realizes smart contract and data verification

Division	Contents
Wallet generating	Generate the original address containing user information into wallet
Node information	Check connected blockchain network
Repository	Create the managing repository for document, data depending on the kind of asset
Registration	Register the data to be created with the original
Verification	Compare has for the verification of data authentication based on registered ledger

shown in Fig. 8. If this is provided through a watermark or QR code, consumers can
check through QR readers.

4   Conclusion

4.1   Summary and Implications of Study

E-document data may be new, but it is already kept by the entity of the transaction—
the agency—that is, it needs to be transferred to the blockchain system for service.
Moreover, it also needs to verify and organize what needs to be provided in order to

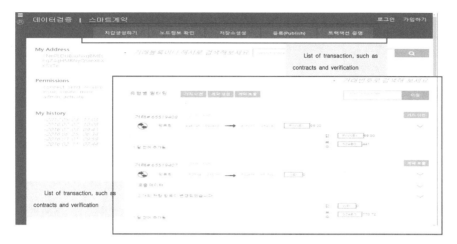

Fig. 7 Blockchain-based electronic transaction system

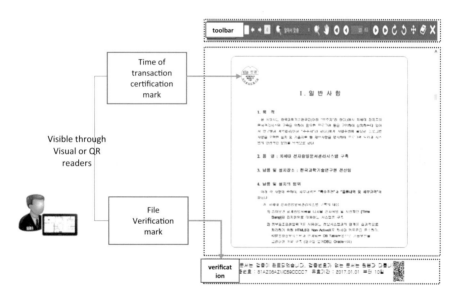

Fig. 8 File viewer

provide a platform for the general consumer. For this, filtering of electronic document files was used to validate electronic documents. Next, a blockchain system was applied to verify the integrity of the forgery problem that could arise when trading electronic documents. Finally, we implemented a pilot system that was configured to be understood by the general consumer when they are provided with electronic document services. However, there are limitations to the verification of electronic documents conducted in this study, and it is necessary to further study these areas in the future.

References

1. Zikratov I, Kuzmin A, Akimenko V, Niculichev V, Yalansky L (2017) Ensuring data integrity using blockchain technology. In: 2017 20th Conference of open innovations association (FRUCT). IEEE, pp 534–539
2. Swan M (2015) Blockchain: Blueprint for a new economy. O'Reilly Media, Inc.
3. Tang H, Shi Y, Dong P (2019) Public blockchain evaluation using entropy and TOPSIS. Expert Syst Appl 117:204–210
4. Oh SY, Cho SH, Han SH, Gim GY (2019, July). A study on the pre-verification of data and the implementation of platform in electronic trade using blockchain. In 2019 20th IEEE/ACIS International Conference on Software Engineering, Artificial Intelligence, Networking and Parallel/Distributed Computing (SNPD) pp 320–330, IEEE
5. Tanwar S, Parekh K, Evans R (2020) Blockchain-based electronic healthcare record system for healthcare 4.0 applications. J Inf Secur Appl 50:102407
6. Lee GY, Kim IH (2019) A study on application of record management system blockchain technology. Korean J Arch Stud 60:317–358
7. Mao D, Hao Z, Wang F, Li H (2019) Novel automatic food trading system using consortium blockchain. Arab J Sci Eng 44(4):3439–3455
8. Morris DZ (2016) Leaderless, blockchain-based venture capital fund raises $100 million, and counting. Fortune (magazine), 05–23
9. Cocco L, Pinna A, Marchesi M (2017) Banking on blockchain: Costs savings thanks to the blockchain technology. Future Internet 9(3):25
10. McGhin T, Choo KKR, Liu CZ, He D (2019) Blockchain in healthcare applications: research challenges and opportunities. J Netw Comput Appl 135:62–75
11. Sarmah SS (2018) Understanding blockchain technology. Comput Sci Eng 8(2):23–29
12. Toyoda K, Mathiopoulos PT, Sasase I, Ohtsuki T (2017) A novel blockchain-based product ownership management system (POMS) for anti-counterfeits in the post supply chain. IEEE Access 5:17465–17477
13. Saberi S, Kouhizadeh M, Sarkis J, Shen L (2019) Blockchain technology and its relationships to sustainable supply chain management. Int J Prod Res 57(7):2117–2135
14. Longo F, Nicoletti L, Padovano A, d'Atri G, Forte M (2019). Blockchain-enabled supply chain: an experimental study. Comput Ind Eng 136:57–69. (Data Source: National Center for Education Statistics, 1998–2007)
15. Nofer M, Gomber P, Hinz O, Schiereck D (2017) Blockchain. Bus Inf Syst Eng 59(3):183–187
16. Ying W, Jia S, Du W (2018) Digital enablement of blockchain: Evidence from HNA group. Int J Inf Manage 39:1–4
17. Gordon WJ, Catalini C (2018) Blockchain technology for healthcare: facilitating the transition to patient-driven interoperability. Comput Struct Biotechnol J 16:224–230
18. Jia B, Zhou T, Li W, Liu Z, Zhang J (2018) A blockchain-based location privacy protection incentive mechanism in crowd sensing networks. Sensors 18(11):3894
19. Meng W, Tischhauser EW, Wang Q, Wang Y, Han J (2018) When intrusion detection meets blockchain technology: a review. IEEE Access 6:10179–10188
20. Xu X, Liu Q, Zhang X, Zhang J, Qi L, Dou W (2019) A blockchain-powered crowdsourcing method with privacy preservation in mobile environment. IEEE Trans Comput Soc Syst 6(6):1407–1419
21. Civelek ME, Özalp A (2018) Blockchain technology and final challenge for paperless foreign trade. Eurasian Acad Sci Eurasian Bus Econ J 15:1–8

22. Ghazali O, Saleh OS (2018) A graduation certificate verification model via utilization of the blockchain technology. J Telecommun Electron Comput Eng (JTEC) 10(3–2):29–34
23. Nizamuddin N, Salah K, Azad MA, Arshad J, Rehman MH (2019) Decentralized document version control using ethereum blockchain and IPFS. Comput Electr Eng 76:183–197
24. Wang J, Wu P, Wang X, Shou W (2017) The outlook of blockchain technology for construction engineering management. Front Eng Manage 67–75. (Data Source: National Center for Education Statistics, 1998–2007)
25. Yang M, Zhu T, Liang K, Zhou W, Deng RH (2019) A blockchain-based location privacy-preserving crowdsensing system. Future Gener Comput Syst 94:408–418

Study on Security and Privacy of E-Government Service

Sanghyun Cho, Saeyong Oh, Hogun Rou, and Gwangyong Gim

Abstract In the big data era, a shift from an existing provider-oriented e-government to a user-centered digital government is required. It is time to implement a consumer-centered digital government that can proactively provide customized services to the public rather than services that meet the public's eye. This study analyzed prior studies and cases of e-government in order to derive the factors influencing the continuous use of e-government services provided by central government departments, local governments and public institutions, and to establish the causal relationship between each factor. Moreover, based on the information system success model and the technology acceptance model, the empirical analysis was used to determine how concerns about quality and privacy of e-government services and security affect usability, user satisfaction, and willingness to use them continuously. This contributed to presenting the policy direction and implications to the e-government in the contemporary paradigm of new IT and the fourth industrial revolution.

Keywords E-government · Quality of e-government services · Security · Privacy · Trust · User satisfaction · IS success model · TAM

S. Cho · S. Oh · H. Rou · G. Gim (✉)
Department of IT Policy and Management, Soongsil University, Seoul, South Korea
e-mail: gygim@ssu.ac.kr

S. Cho
e-mail: cshyms74@naver.com

S. Oh
e-mail: ohsy@satu.co.kr

H. Rou
e-mail: hgr1203@soongsil.ac.kr

© The Author(s), under exclusive license to Springer Nature Switzerland AG 2021   67
J. Kim and R. Lee (eds.), *Data Science and Digital Transformation in the Fourth Industrial Revolution*, Studies in Computational Intelligence 929,
https://doi.org/10.1007/978-3-030-64769-8_6

1   Introduction

With the development of ICT (Information Communication Technology) and the advent of web browsers, the environment in which anybody can use internet easily has been created. As a result, people's Internet access has begun to spread rapidly. These social changes have brought the implementation of e-government, which uses ICT to innovate government processes. In particular, the government promoted innovation of administrative services that applied ICT to government affairs to enhance administrative efficiency and transparency, and drastically improved the quality of online services to the public, providing the public with easy and convenient access and usage of various information and services provided by the government, local governments, public organizations and financial institutions. This has brought new changes throughout the society as well as in daily life [1].

Accordingly, services that can provide high-quality services and a safe environment to the level the public wants and meet various demands are now becoming a necessity, not an option of the e-government. In particular, since the ultimate goal of e-government is to promote service use and provide safe service rather than to build a system, it is very important to pinpoint the problems and solutions of users' needs, privacy and service safety [1].

Furthermore, it is important for the e-government to seek ways to restore public trust and continuously develop e-government services through in-depth analysis of problems and service issues at the service site for those who have used the e-government service, as it is a very important task for the e-government to raise the public's intention to use e-government services through continuous promotion to the government and provision of various services.

Therefore, this research is meaningful in deriving factors that influence user satisfaction with e-government services and the intention of continuous use of services in order to help respond to changes in ICT environment and social structure such as information technology development, distribution of smartphones, IoT, and artificial intelligence (AI) in Korea, and implement e-government that can meet the needs of the people. To accomplish this, the research first analyzed existing research on e-government to find success factors that can contribute to improving service quality, and quality (service quality, system quality, information quality). The research hopes to identify the impact of causal risk (personal information protection, security) factors on user value (usability, usability, trust) and user satisfaction to study ways to continuously use e-government services and suggest implications.

2   Theoretical Background

2.1   E-Government

The concept of e-government is an extension of the concept that first emerged from the e-Bank Service mentioned in the 1993 National Performance Review [2], which was promoted by the Clinton administration in the United States. It refers to the implementation of a government using information technology to deliver administrative and public services in a more convenient, customer-oriented and innovative manner. It also refers to the delivery of necessary administrative and civil services or administration through electronic means such as the Internet and smartphones. However, the concept of e-government varies depending on the country pursuing e-government, academia and consulting institutions working on it, and there is no agreed definition yet [1].

The concept and definition of e-government are changing little by little, as e-government can be viewed differently depending on the scope of application and interpretation from what perspective it interprets as the development of information and communication technology like Cloud, artificial intelligence(AI) and changes in the social environment.

It is very important for e-government to build trust with all stakeholders, and if the people's trust in e-government increases, they will be able to secure support from the people. If trust is lowered, the people will be discredited. In this study, among the factors affecting e-government trust, the focus is on personal information protection and security, and the factors affecting the service users and the e-government that is the subject of trust are to be verified with empirical evidence [3].

First, customer orientation is the most important conceptual characteristic in service quality, one of the factors affecting e-government. The service provided by e-government should be evaluated by the people and the satisfaction of the service should be measured from the user's point of view. Most of the preceding studies on e-government service quality are based on theoretical background on factors affecting service acceptance, such as service utilization, user satisfaction, and acceptance from a consumer-centered perspective. These studies began with the recognition of problems with relatively low utilization of e-government services despite the high level of e-government deployment [4].

In an e-government service quality scale study considering relevance, Lee [5] confirmed, through an empirical analysis, that related quality has a significant impact on individual efficiency and civic satisfaction in e-government service by adding new quality to e-government service quality.

2.2   Factors for Choosing E-Government

Security in e-government, which provides services based on the Internet environment, refers to an unauthorized system access, risk and information protection over the Internet. Due to the importance of information, various studies are being conducted in the areas of data security and privacy [1].

Prior research on personal information focuses on identifying the causal relationship of factors that affect users' privacy concerns and intentions to provide personal information [6], and explaining privacy-related attitudes, intentions and behaviors centered on users' risks and benefits from a technical and tool perspective. For e-government, studies on privacy issues have mainly been addressed in legal and institutional aspects [7], and prior studies on security in e-government services have also been mostly case studies that apply security practices to information protection management systems [1].

In addition, looking at the preceding research on user satisfaction and acceptance of mobile e-government services, Han identified factors that affect the acceptance of mobile e-government services and empirically analyzed whether each factor affects user satisfaction and willingness to reuse, suggesting ways to increase and activate mobile e-government services [8]. System quality, service quality, information quality, related quality, and public quality were derived as independent variables, and user satisfaction and intention to reuse were selected as factors for usability. Usability and acceptance were parameters to identify causal relationship between each factor. As a result, system quality, service quality, and related quality among quality factors have been proven to have positive effects on usability, and information quality has been proven to have a positive effect on causal relationship to usability [9]. Also, the casual relationship between the user value (usability, usefulness) factors and acceptance intention (user satisfaction, intention of reuse) factors have been proven [10]. This proved that the technology acceptance model and the information system success model are suitable models to describe mobile e-government services [11].

As proven, e-government is the process of selecting services that provide existing administrative services in new and innovative ways. The question of what factors people are considering in choosing new ways of service for the proliferation of e-government services, activation of use, and search for ways to improve services is a very important issue [1].

2.3   Information System Success Model (IS Success Model)

This study discussed the impact on the continuous use of e-government services from the perspective of the information system success model on the reliability of e-government, users' satisfaction with the service, and the intention of use due to risk factors such as service quality, privacy, and security.

DeLone and McLean analyzed key research papers related to information system evaluation to review the requirements and corresponding indicators for information system performance [12]. The study reviewed system quality, information quality, usage, satisfaction, individual impact, and organizational performance. Davis also argued through various literature surveys that perceived usefulness and perceived use form a causal relationship with the user's attitude, and that this attitude affects the position of action and affects actual use [13]. Subsequently, DeLone and McLean [14] presented a modified model that added a service quality area as a leading quality factor for user satisfaction. Jin et al. [15] analyzed how mobile payment systems affect personal benefits based on the information system success model and confirmed that the quality of the systems and information affects use and user satisfaction, and that use and user satisfaction are the main factors that have a positive impact on personal benefits.

2.4   Technology Acceptance Model (TAM)

Technology acceptance model, a model developed based on Davis's existing rational theory of behavior, is designed to predict users' acceptance of new technologies [15]. It is set to perceived usefulness and perceived ease of use as factors that affect acceptance, which affect attitudes, affect the intent of action to use them, and explains the human and human structures in which the intent of action affects actual use [1].

The technology acceptance model is recognized as a concise and very descriptive model in explaining consumers' acceptance and usage behavior of information technology. It has been treated as a major model in many empirical studies on the intent to accept new technologies since the 1990s.

In addition, because the technology acceptance model provides an explanation of the determinants of computer acceptance and is comprehensive enough to explain the behavior of users across computer technology and user populations, studies of what factors influence the adoption of new IT technologies that are emerging with the advent of the Fourth Industrial Revolution [1].

3   Design of Study

3.1   Research Model

This study will construct the study model by referring to the preceding study and discuss the results of the analysis. Based on empirically proven variables, quality factors (service quality, system quality, information quality) and risk factors (personal information, security) are set as independent variables. The conclusions are drawn

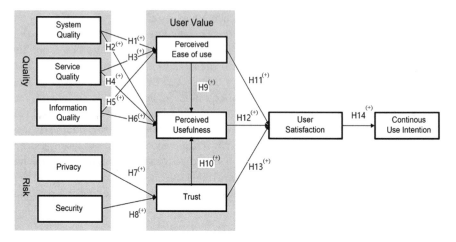

Fig. 1 Research model and hypothesis

based on the research model conducted in advance of establishing user satisfaction and continuous use intent for e-government services by parameters (intentional usability, perceived usefulness, trust), and dependent variables [1].

Based on these variables, a research model was presented, as shown in Fig. 1, to demonstrate the causal relationship between the factors, focusing on how quality and risk factors affect the user's value factors, how useful and easy to use, and how users would be satisfied with the service and would intend to continue to use the e-government service if they were satisfied [1].

The categories of variables being considered in this study are quality factors (service quality, system quality, information quality) and risk factors (personal information reporting, security) as independent variables. Recognition factors are perceived usability, usability, and trust are parameters, and user satisfaction and continuous use are acceptance factors set as dependent variables. The following hypotheses were established by reference to the results of the preceding study [1] to determine the relationship between each variable on how these influencing factors affect the acceptability factors through the cognitive factors (Table 1).

4 Empirical Analysis

4.1 Data Collection

The survey used in the study was composed of 60 questions, referring to the study [1]. Each question was measured on a seven-point scale of the recertification, which measures respondents' perception of e-government services. IBM SPSS Statistics22 and Smart PLS 2.0 were used as tools for analysis.

Table 1 Variable's operational definition

Factor (measured variable)	Operational definition	References
Service quality (SerQ)	The degree of perceived value of the user for the service	[1, 16–20]
System quality (SysQ)	The degree of perceived value of the user to the system	
Information quality (InfQ)	The degree of perceived value of the user for the information	
Privacy (Pri)	The degree of perceived value for personal information management, infringement, change, leakage, etc.	
Security (Sec)	The degree of perceived value for service safety	
Perceived usefulness (PU)	The degree to which users perceive that they are useful for their purpose	
Perceived ease of use (PEU)	The degree to which users perceive that the service is easy and convenient	
Trust (Tru)	The degree to which believe trust the service	
User satisfaction (US)	The degree of overall user satisfaction after using the service	
Continuous use intention (ICU)	Intended to continue to use the service in the future	

First, frequency analysis was performed to examine the demographic characteristics and generalities of the survey respondents. The collected samples were collected from a total of 318 people using the questionnaire method for those who have used e-government services. The data were analyzed in 284 parts, excluding those who had double misses or answered insincerely (34 surveys). In terms of demographic characteristics, 121 males (58.17%) and 87 females (41.83%) were surveyed. In terms of age distribution, people in their 40s accounted for 34.62%, followed by those in their 30s (32.69%) and those in their 20s (20.19%).

4.2 Exploratory Factor Analysis

An exploratory factor analysis (EFA) was performed to verify the reliability and validity of detailed measurement items, and a confirmatory factor analysis (CFA) was performed to analyze model suitability and verify reliability and validity. The Cronbach-Alpha coefficient was used to verify reliability. Cronbach-Alpha is an indication of whether the measured items consist of the same components based on the mean correlation, and for reliability analysis, Nunally set the reliability of the concept of the configuration at 0.7 and determined that a value of 0.7 or higher would not be a problem with the reliability of the target variable [21]. For the purpose of validation, the degree of association of embedded factors of the variables to be observed was explored in an exploratory factor analysis (EFA) to identify the structure between factors. A factor analysis indicates that an absolute value of factor loading is 0.4 or higher is considered a significant variable, and a value greater than 0.5 can be determined as an important variable [1].

According to the results of an exploratory factor analysis (EFA), reliability analysis was performed by deleting SysQ2, PU5, PEU 1&5, SerQ 2&3, Pri 4&5, InQ 4&5, and Tru 5. Cronbach-alpha values of all variables were secured above 0.7.

4.3   Confirmatory Factor Analysis

4.3.1   Appropriateness Analysis of Structural Model

The measurement variables extracted through EFA (Exploration Factor Analysis) were performed using Smart PLS 2.0. Analysis of conformity throughout the structural model can be verified by indicators such as Redundancy value, R^2 value, and goodness of fit [22].

Looking at the criteria for each metric, the redundancy index is an indicator of the statistical estimate of the structural model and is evaluated as having the suitability of the structural model when the redundancy value is positive. Cohen was able to verify the model's suitability by the R^2 value of the endogenous variables, divided into 'high' if the R^2 value was 0.26 or higher, 'middle' if the R^2 value was greater than 0.13–0.26 and 'low' if they were less than 0.02 and more [23]. Tenenhaus et al. [24] divided the overall suitability into 'high' if the magnitude of the fit is 0.36 or higher, 'middle' if the fit is greater than 0.25 and less than 0.36 and 'low' if it is between 0.1 and 0.25 and less.

In this study, as shown in Table 2, the suitability of the measurement model can be explanatory because all of the redundancy values have positive values. The R^2

Table 2 Reliability analysis of confirmative factor analysis

	AVE	Composite reliability	R^2	Cronbach's Alpha	Communality	Redundancy
Pri	0.857	0.947	☐	0.917	0.857	☐
Sec	0.866	0.970	☐	0.961	0.866	☐
US	0.782	0.947	0.715	0.930	0.783	0.367
SerQ	0.785	0.916	☐	0.863	0.785	☐
SysQ	0.741	0.920	☐	0.884	0.741	☐
Tru	0.850	0.958	0.643	0.941	0.850	0.348
InQ	0.828	0.935	☐	0.896	0.828	☐
PEU	0.817	0.931	0.616	0.888	0.817	0.251
PU	0.814	0.946	0.677	0.924	0.814	0.197
ICU	0.834	0.962	0.453	0.950	0.834	0.377

value is also 0.26 or higher, and the results of the calculation of the adequacy of the structural model all exceed the criteria and are considered highly suitable. The overall goodness-of-fit analysis showed that the value was 0.712, indicating a value greater than 0.36 and higher, indicating a higher model fit.

As a result of checking the suitability of the structural model in this study, as shown in Table 2, the measurement results for redundancy, R^2, and overall suitability all exceeded the criteria, indicating that the model is descriptive [1].

4.3.2 Reliability Analysis

The suitability of the previous structural model was shown to be descriptive, and a Confirmation Factor Analysis (CFA) was conducted to establish the relationship between the variables. Typically, average variance extractions (AVE: Average Variance Extracted) and composite reliability (CR: Composite Reliability) are used to verify reliability. Fornell & Larker stated that reliability is ensured only when the mean of variance (AVE) extracted on a reliability basis is greater than 0.5 and the composite reliability (CR) is greater than 0.7. In this study, the mean variance extract (AVE) of all variables is greater than 0.5 and the composite reliability (CR) is greater than 0.7 [25]. and all values are above the reference value as shown in Table 2 [1].

4.3.3 Discriminant Validity Analysis

Because the model's suitability was explained earlier and reliability was ensured by the positive factor analysis, a discriminant feasibility analysis was conducted to confirm the validity of the model. Fornell & Larker assesses that the mean variance extract (AVE) obtained between each potential variable is justified when the square of the correlation, or square root, appears to be greater than the coefficient of determination [25]. As analyzed in Table 3 in this study, the highest correlation in the correlation matrix of the potential variables is that the coefficient of determination squared by the user satisfaction and confidence (0.801) is 0.642 (0.801800.801), so the average variance extract (AVE) obtained between all potential variables is higher than the coefficient of determination. Thus, it can be considered that it is comparable [1].

4.3.4 Hypothesis Verification

In this study, path analysis was performed to verify each hypothesis using Smart PLS2.0. The hypothesis verification of the structural model can be verified by the t-values provided by bootstraping of the PLS. It can be considered statistically significant at a significant level of 0.01 if the absolute value of the t-values is greater than 1.65, a significant level of 0.05 if greater than 1.96, and a significant level greater than 2.33, a significant level of 0.02 and 2.58. As a result of the hypothesis test, all

Table 3 Result of discriminant validity

	Pri	Sec	US	SerQ	SysQ	Tru	InQ	PEU	PU	ICU
Pri	0.857	□	□	□	□	□	□	□	□	□
Sec	0.798	0.866	□	□	□	□	□	□	□	□
US	0.703	0.765	0.782	□	□	□	□	□	□	□
SerQ	0.644	0.637	0.690	0.785	□	□	□	□	□	□
SysQ	0.694	0.734	0.707	0.732	0.741	□	□	□	□	□
Trus	0.747	0.771	0.801	0.724	0.681	0.850	□	□	□	□
InQ	0.677	0.736	0.725	0.709	0.778	0.741	0.828	□	□	□
PEU	0.657	0.681	0.769	0.697	0.713	0.772	0.728	0.817	□	□
PU	0.692	0.742	0.731	0.711	0.671	0.767	0.743	0.685	0.814	□
ICU	0.602	0.645	0.673	0.652	0.584	0.707	0.647	0.562	0.775	0.834

but two of the 14 hypotheses based on the study model had a significant impact and 12 hypotheses were adopted.

Among the two rejected hypotheses, such as the research hypothesis claimed by Cho et al. [1], the relationship between system quality and perceived usefulness does not appear to affect the usefulness of e-government services. This is because the system must be professionally designed for e-government services, and the securing of system stability (down-error, etc.), and the prompt resolution of problems are perceived as a natural service that must be performed in the operation of the information system. In addition, the relationship between perceived usability and perceived usefulness is not a useful service for users, although it is easily and conveniently used at anytime and anywhere. if they do not have the necessary information or are not helpful to their daily lives.

The results of the path analysis and the adoption of each hypothesis in this study are shown in Table 4.

5 Conclusions

5.1 Summary of Research Result and Implications

In this study, prior research and case studies on e-government were analyzed to determine the correlation between the quality factors (service quality, system quality, information quality), risk factors (personal information protection, security), and user value factors (understanding usability, perceived usefulness, trust) of e-government services. Moreover, the study aimed to establish the effect of each factor on user satisfaction and the degree of continuous use of e-government service. Most of the hypotheses have been adopted as in previous studies, but the parts that differ from previous studies are as follows.

Table 4 Hypothesis verification result

Hypothesis	Path factor	T-value	Result
Personal information → trust	0.362	3.632***	Adopted
Security → trust	0.482	5.366***	Adopted
User satisfaction → continuous use intention	0.673	14.909***	Adopted
Service quality → perceived easiness	0.275	2.520*	Adopted
Service quality → perceived usefulness	0.198	2.421*	Adopted
System quality → Perceived easiness	0.244	2.128*	Adopted
System quality → perceived usefulness	0.032	0.316	Rejected
Trust → user satisfaction	0.389	3.701***	Adopted
Trust → perceived usefulness	0.365	3.096**	Adopted
Information quality → Perceived easiness	0.343	2.434*	Adopted
Information quality → perceived usefulness	0.278	2.362*	Adopted
Perceived easiness → user satisfaction	0.325	3.961***	Adopted
Perceived easiness → perceived usefulness	0.040	0.382	Rejected
Perceived usefulness → user satisfaction	0.210	2.468*	Adopted

$*p < 0.05, **p < 0.01, ***p < 0.001$

First, System quality does not affect perceived usefulness. E-government ser-vices should provide convenient access to services that users need anytime, anywhere, and providing reliable, up-to-date information to service users through continuous updates is an important factor in determining the usefulness of e-government services. However, in the case of the system, reflecting the new technology, securing stability (e.g. down-error), and providing prompt resolution in the event of a problem does not affect the degree of user convenience. Rather, it is perceived as a natural service that must be performed in the operation of the information system.

Second, the relationship between perceived usability and perceived usefulness was found to have no effect on perceived usefulness. Apart from easily and conveniently using the desired service by accessing the service anytime, anywhere through various means such as web and mobile, this service can be interpreted as not useful to users if there is no information they need or if the service and information provided do not help them in their daily lives.

Based on prior research, this study can be meaningful in that the factors influencing the continuous acceptability of e-government services were theoretically present-ed. In addition, the operational definitions, reliability, and validity of the independent variables, parameters, and dependent variables presented in the study model provided a theoretical foundation to help empirical research similar to this study in the future.

5.2   Limitations of the Study and Direct of Future Study

This paper would like to clarify the limitations of the study that have not been identified in this research and suggest future research directions to supplement them.

First, the investigation on the perception of e-government services on diverse subjects was insufficient. The survey targets were limited to those who had experience using the services provided by the e-government, and no surveys were conducted on those aged 60 or older. Therefore, a detailed statistical analysis is needed by expanding the survey respondents' perception of e-government by those who have no experience in using e-government services or who do not receive service benefits due to the digital divide. This will be done by securing many samples.

Second, there was a lack of study on the various factors that affect e-government trust. Although this study confirmed that trust in e-government is a very important factor in user satisfaction and continuous use of e-government services, the research was conducted only in terms of privacy and security. Thus, more detailed research is needed to present various factors affecting trust in order to accurately investigate how reliable the e-government is and what factors affect e-government trust.

Finally, since new technologies such as IoT, artificial intelligence (AI), and block chain should be applied and policies and laws should be supported to implement intelligent e-government suitable for the era of the Fourth Industrial Revolution, it is necessary to study what kind of perceptions that users who use e-government services have an awareness of e-government-related systems and how these perceptions affect the intention of continuous use of e-government services.

References

1. Cho SH, Oh SY, Rou HG, Gim GY (2019) A study on the factors affecting the continuous use of e-government services-focused on privacy and security concerns. In: 2019 20th IEEE/ACIS international conference on software engineering, artificial intelligence, networking and parallel/distributed computing (SNPD). IEEE, pp 351–361
2. National Intelligence Service, Ministry of Public Administration and Security, Ministry of Knowledge Economy, & Korea Communications Commission (2009) National Information Protection White Paper. Seoul, Korea
3. Aaker D (2003) The power of the branded differentiator. MIT Sloan Manage Rev 45(1):83; Al-Omari H, Al-Omari A (2006) Building an e-Government e-trust infrastructure. Am J Appl Sci 3(11):2122–2130
4. Kwon YM (2014) A study on the effect of service quality on service satisfaction and government trust in the government 3.0 era—focusing on the information gap between users in the metropolitan area. Doctoral dissertation. Available from RISS Dissertation and Theses database. (No. 13417330)
5. Lee CE (2008) The effect of relationship quality on citizen satisfaction with electronic government services. Doctoral dissertation. Available from RISS Dissertation and Theses database. (No. 11274561)
6. Lee HG, Lee SH (2009) Analysis of prior studies for elucidating future directions of information privacy: focusing on information privacy concerns online. Informatization Policy 16(2):3–16

7. Choi BM, Park MJ, Chae SM (2015) A study on factors affecting personal information technology acceptance behavior. Inf Syst Rev 17(3):77–94
8. Han KH (2012) An empirical study on the influencing factors of intention to adoption of mobile e-government service. Doctoral dissertation. Available from RISS Dissertation and Theses database. (No. 12683689)
9. Elliman T (2006) Generating citizen trust in e-government using a trust verification agent: a research note. In: European and mediterranean conference on information systems
10. Warren SD, Brandeis LD (1890) The right to privacy. Harvard Law Rev 193–220
11. Kim BS, Lee J, Kim KK (2006) The impact of perceived trust, risk, usability, and convenience on intention to use e-government service: focused on online application and service. Informatization Policy 13(4):186–202
12. DeLone WH, McLean ER (1992) Information systems success: the quest for the dependent variable. Inf Syst Res 3(1):60–95
13. Davis FD (1989) Perceived usefulness, perceived ease of use, and user acceptance of information technology. MIS Q 319–340
14. Delone WH, McLean ER (2003) The DeLone and McLean model of information systems success: a ten-year update. J Manage Inf Syst 19(4):9–30
15. Jin JS, Kim HM, Park JS (2017) A study on behavior in using fin-tech based on life style types. J Inf Technol Serv 16(1):119–138
16. Pitt LF, Watson RT, Kavan CB (1995) Service quality: a measure of information systems effectiveness. MIS Q 173–187
17. In KY (2016) A study of the effect of service quality to the continuous usage of e-government services. Master thesis. Available from RISS Dissertation and Theses database. (No. 14059381)
18. Davis FD (1985) A technology acceptance model for empirically testing new end-user information systems: theory and results. Doctoral dissertation, Massachusetts Institute of Technology
19. Kim BS (2018) An empirical study on factors affecting the operational performance of the war game system of the ROK air force: focused on theater-level exercise model. Doctoral dissertation. Available from RISS Dissertation and Theses database. (No. 14730384)
20. Weinstein L, Neumann PG (2000) Internet risks. Commun ACM 43(5):144–144
21. Nunnally JC (1978) Psychometric theory. McGraw-Hill, New York
22. Chin WW (1998b) Issues and opinion on structural equation modeling. Mis Q 22(1)
23. Cohen J (1988) Statistical power analysis for the behavioral sciences. Psychology Press
24. Tenenhaus M, Vinzi VE, Chatelin YM, Lauro C (2005) PLS path modeling. Comput Stat Data Anal 48(1):159–205
25. Fornell C, Larcker DF (1981) Evaluating structural equation models with unobservable variables and measurement error. J Mark Res 18(1):39–50

Applying Hofstede's Culture Theory in the Comparison Between Vietnam and Korean E-Government Adoption

Hung-Trong Van, Simon Gim, Euntaek Lim, and Thi-Thanh-Thao Vo

Abstract Due to the rapid development of information technology (IT), E-government has become a popular issue all over the world. E-government system has also become the standard to comparing the technical development background of many countries around the world. Keeping up with the trend, Vietnam Government has been focusing on developing E-Government system for years and E-government has become one of the most important priorities. Compared to Vietnam, which is at the beginning stage, Korean E-Government is one of the top systems in the E-Government World Ranking. Based on Hofstede's National Culture theory, this study finds the difference of E-government system/service between Vietnam and Korea under the evaluations of citizens.

Keywords E-government · Korea · Vietnam · Difference · National culture theory · TAM · IS success model

1 Introduction

Many policy makers and politicians in both developing and developed countries have been attracted to use E-government as a way to develop their countries. Unlike

H.-T. Van (✉) · T.-T.-T. Vo
Faculty of Digital Economy & E-Commerce, Vietnam–Korea University of Information and Communication Technology, Danang, Vietnam
e-mail: vhtrong@vku.udn.vn

T.-T.-T. Vo
e-mail: thaovo90dn@gmail.com

S. Gim
SNS Marketing Research Institute, Soongsil University, Seoul, Korea
e-mail: simongim93@gmail.com

E. Lim
Graduate School of Business Administration, Soongsil University, Seoul, Korea
e-mail: iet030507@gmail.com

© The Author(s), under exclusive license to Springer Nature Switzerland AG 2021
J. Kim and R. Lee (eds.), *Data Science and Digital Transformation in the Fourth Industrial Revolution*, Studies in Computational Intelligence 929,
https://doi.org/10.1007/978-3-030-64769-8_7

other systems which are still in controversy regarding privatization and globalization [1], E-government is recognized as an effective solution which has been quickly developing and spreading across the world. To establish this E-government, many countries have been investing a lot of money and political commitment [1]. In many developing countries such as Indonesia, India, and Vietnam, E-government initiatives have been collected to create the opportunities for ICT and economic development [2]. E-government system is viewed as an effective tool to reduce corruption and enhance economic development. Not only that, it can improve the services' quality for their citizens [3].

Although E-government systems have been used in both developing and developed countries, not all E-government system's diffusion has gained expected indicated outcomes [4]. In case of developing countries, the diffusion process remains slow due to the incessant shortage of resources [5]. Almost all the developing countries suffer from lack of finance due to weak national information infrastructure and insufficient knowledge or skill to develop effective and appropriate strategies for E-government. This research focuses on finding the factors affecting the intention to use E-government by comparing Korea and Vietnam citizens' behavior based on TAM and IS success Model. Furthermore, the Hofstede's theory on national culture is used to find out the difference between Korea and Vietnam.

2 Literature Review

2.1 E-Government in Korea

Korean E-government was first implemented in 1978 with ambition to become the E-government with the highest developing E-government index in the world. To start with, significant policies and strategies were established by the Korean government to promote the growth of government service under the aid of ICT. After 40 years of implementing E-government, there has been many noticeable remarks made in Korea. First, government's agencies and departments were computerized and connected to the Internet in the late 1970s. Secondly, several important projects were implemented and deployed including the Five National Computer Network and Comprehensive Plan for Korea Information Infrastructure Establishment. In 2001, E-government development strategies which concentrated on contributions and diversification of delivery channel in public services were established [2, 6]. Through numbers of concentrated projects, Korean government built a high-speed communication network and proper storing systems to put government record in digital formats. These formats provided solid foundation for E-government implementation in Korea [6].

2.2   E-Government in Vietnam

Vietnam government has been providing E-government services to citizens for many years [2]. In most developing countries, E-government is recognized as a potential tool to improve government institutions' capacities and offer chances to better solve problems in the public administration [7]. Nevertheless, as a developing country with low level of IT adoption and economic development, Vietnam has numerous issues in applying and implementing E-government services/system. Therefore, the current statement of Vietnam E-government is still far below the government's expectation. It does not attract significant concern from citizens as well as business areas [8]. E-government services are still in its inception with citizens and businesses, and the outcome of E-citizenship dimension is even more limited [9].

2.3   Information System's Theories

Technology Acceptance Model (TAM) of Davis and the original and updated IS success models by DeLone and McLean are considered as one of the most reputable models that have been extensively used for predicting and explaining underlying factors. TAM theorizes that the behavioral intentions of one person are decided by two belief constructs as perceived usefulness and perceived ease of use. The aims of IS success model are defining and clarifying the relationships among six of the most important dimensions which evaluate any information systems.

2.4   Hofstede's National Culture Theory

National culture theory of Hofstede is a well-known framework for any cross-cultural research. In this theory, the effects of national culture on many research areas are described based on the values of members in that society and how the value links with human behavior. This is done by using a structure which is acquired from factor data analysis [2]. According to Hofstede, the differences of national cultures are identified based on six primary dimensions. These dimensions are power distance, collectivism and individualism, uncertainty avoidance, masculinity and femininity, long-term orientation and short-term orientation, and indulgence and restraint.

3   Research Model and Hypotheses

Considering the IS success model, TAM, national culture theory, and other researches about E-government, the research model was proposed in Fig. 1 [2].

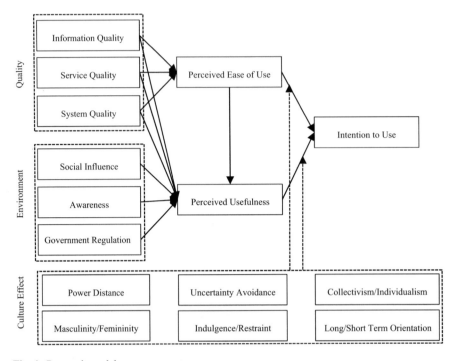

Fig. 1 Research model

According to Delone and McLean, information quality is identified with how much the given information fits the requirements of client [10]. In E-government literature, information quality is characterized with respect to some subsets or some assortments of the accompanying traits: relevancy, validity (accuracy), completeness (breadth and comprehensiveness), understandability, accessibility, reliability (dependability), timeliness (currency), concise presentation, authority, objectivity, security, appropriate amount (quantity), perceived value and freedom from error [2]. System quality is characterized as the information system's quality processing that incorporates data and software parts. In the context of E-government services, government sites ought to give prompt, simple, and easy information approach regarding technology [11]. This gives a definite influence on perceived usefulness and perceived ease of use because of improved adequacy and productivity [12]. Service quality is related to the degree to which the services are delivered to address customers' needs and in relation to the entity managing the support of the system [10], particularly in the context of E-government, service quality plays an important role [12]. Moreover, users in every service encounter normally expect a satisfactory level of service quality when engaging in any E-government services [12].

In accordance with Venkatesh and Davis, social influence is one of the key elements in discovering usage behavior as well as users' acceptance of any novel

IT [13]. Taylor and Todd pointed out that social influence refers to the subjective norms such as peers influence, superior influence as well as the opinion from other people [14]. Additionally, the approving or disapproving perceptions of people towards the E-government usage might be affected by the perceptions of their family, colleagues, or acquaintances influence [2]. Awareness is related to an understanding of others' behaviors, from that to provide a framework for your own behavior [15]. In E-government context, using mass media and carrying out workshops are one of methods for introducing the concept as well as encourage citizens to use E-government services in their daily life. Government regulation refers to government laws, regulations, and policies that any associations must comply with it. Regulation environment is considered as important factor in IT innovation adoption [16]. In the context of E-government, it is necessary to have new policies and regulations for developing E-government assimilation. If government doesn't offer a suitable regulatory framework for using E-government, associations might discourage or corrupt in using it [2].

Perceived usefulness refers to the assessment of users regarding task-oriented outcomes. It includes information regarding the extrinsic of the task such as how IT could help to enhance task efficiency and effectiveness [17]. Perceived usefulness is seen as a significant factor in determining the adoption of innovations [18]. According to Kumar, the adoption of E-government is affected directly by the extend of perceived the usefulness [19]. Perceived ease of use is defined as the extent to which one person has the belief that it would be effortless in using a particular system [20].

Power distance refers to the extent to which the unequal power distribution in organizations is accepted by the less powerful members. People have inclination to accept and finish their tasks as well as duties which is assigned to them by their bosses [21]. In high power distance society, E-government activities as well as IT adoption could be perceived as a threat to the hierarchy [22]. According to Geert and Jan, uncertainty avoidance is related to how much obscure or dubious circumstances danger individuals in a society [23]. In simple way, uncertainty avoidance is the extent that one society's citizens feel intimidated and try to avoid dubious circumstances by ignoring abnormal ideas and behaviors, and designing formal regulations and policies [24]. In high uncertainty avoidance society, citizens usually feel threaten by unstructured, uncertain, and ambiguity circumstances [25]. Collectivism is related to a society in that the connection among people are close; everybody has belief to take care themselves as well as people around them [26]. In individualism cultures, people believe in their decisions and have more self-oriented characteristics. Masculinity, in contrast to femininity, is defined as a culture where there is a clear separation between the emotional roles between the two genders. According to Hofstede and his partners, a higher masculinity society performs a competitive, productive, and assertive centered culture [26]. Conversely, femininity culture is related to a more solidarity, equality, consensus seeking social relationships, and concerning centered culture [27]. In high masculinity society, people are more interested in the usefulness of given technology regardless of whether it is easy or difficult to use [28]. Long-term orientation culture is differentiated by values [26]. Conversely, short-term orientation culture has belief in quick outcomes and relies upon regard for the past focusing on steadiness and custom [29]. In long term orientation society, the connected values are persistence and carefulness. Meanwhile, the associated values in short-term orientation culture are fulfilment with social commitments and respect to customs [2]. Indulgence is identified with a culture that moderately free delight of elemental and natural human ambitions associated with having good times and appreciating life are permitted [26]. Moreover, restraint refers to a culture that the needs' gratification is controlled and directed by severe social standards' ways [26]. Society with high indulgence index does not apply strict social norms to the fulfilment of people about basic needs and desires. Accordingly, people can utilize new technologies when they want it. Conversely, in restraint society, individuals have the constraint on the adoption with new innovations, adopting even their life [26]. The hypotheses are proposed:

H1: Information Quality (InfQ) has positive influence on Perceived Usefulness (PU)

H2: Information Quality has positive influence on Perceived Ease of Use (PEU)

H3: System Quality (SysQ) has positive influence on Perceived Usefulness

H4: System Quality has positive influence on Perceived Ease of Use

H5: Service Quality (SerQ) has positive influence on Perceived Usefulness

H6: Service Quality has positive influence on Perceived Ease of Use

H7: Social Influence (SI) has positive influence on Perceived Usefulness

H8: Awareness (AW) has positive influence on Perceived Usefulness

H9: Government Regulation (GR) has positive influence on Perceived Usefulness

H10: Perceived Ease of Use has positive influence on the Perceived Usefulness

H11: Perceived Ease of Use has positive influence on the Intention to Use E-government (In)

H12: Perceived usefulness has positive influence on the Intention to use E-government

H13: Power Distance (PD) moderates the relationship between Perceived ease of use and Intention to use E-government

H14: Power Distance moderates the relationship between Perceived usefulness and Intention to use E-government

H15: Uncertainty Avoidance (UA) moderates the relationship between Perceived ease of use and Intention to use E-government

H16: Uncertainty Avoidance moderates the relationship between Perceived usefulness and Intention to use E-government

H17: Collectivism/Individualism (CO) moderates the relationship between Perceived ease of use and Intention to use E-government

H18: Collectivism/Individualism moderates the relationship between Perceived usefulness and Intention to use E-government

H19: Masculinity/Femininity (MA) moderates the relationship between Perceived ease of use and Intention to use E-government

H20: Masculinity/Femininity moderates the relationship between Perceived usefulness and Intention to use E-government

H21: Long Term Orientation/Short Term Orientation (LT) moderates the relationship between Perceived ease of use and Intention to use E-government
H22: Long Term Orientation/Short Term Orientation moderates the relationship between Perceived usefulness and Intention to use E-government
H23: Indulgence/Restraint (ID) moderates the relationship between Perceived ease of use and Intention to use E-government
H24: Indulgence/Restraint moderates the relationship between Perceived usefulness and Intention to use E-government

4 Data Analysis

The research model was evaluated by using a designed questionnaire to collect data. The instruments were measured based on 7-point Likers-scales from strongly disagree (1) to strongly agree (7). The survey was carried out for 30 days from September 2015. 462 appropriate responses out of 480 returned responses were used for data analysis giving 96.25% of response rate.

Table 1 shows the results for convergent validity testing including AVE and CR. If AVE is greater than 0.50 and CR greater than 0.7 and CR is over AVE of the total variance, convergent validity is established [2] and the convergent validities of all 9 factors were confirmed.

In Table 2, it can be seen that service quality does not have effect to the perceived usefulness of E-government and perceived ease of use does not affect to the intention to use E-government with the p-value higher than 0.1 [2].

According to Table 3, in case of Korea E-government, the relationship between service quality and perceived usefulness as well as the relationship between system quality and perceived usefulness are rejected with p-values higher than 0.1 [2].

According to Table 4, in case of Vietnam E-government, government regulation and awareness do not have effect on the perceived usefulness of E-government with p-values higher than 0.1 [2].

In Table 5, it shows that power distance cannot moderate the relationship between perceived ease of use and intention to use in the comparison between Korea and Vietnam because the p-value are higher than 0.1. However, power distance can moderate the relationship between perceived usefulness and intention to use E-government in the comparison between Korea and Vietnam when the p-value is less than 0.01 [2]. Uncertainty avoidance cannot moderate the relationship between perceived ease of use and intention to use as well as perceived usefulness and intention to use E-government in the comparison between Korea and Vietnam with p-values

Table 1 CR and AVE

	AW	GR	PU	SI	InfQ	SysQ	PEU	SerQ	In
CR	0.865	0.92	0.88	0.896	0.917	0.908	0.917	0.799	0.899
AVE	0.682	0.741	0.709	0.684	0.735	0.711	0.786	0.572	0.748

Table 2 Structural paths assessment and hypothesis test

	Estimate	S.E.	C.R.	p	Label
PEU <--- InfQ	0.307	0.060	40.824	***	Supported
PEU <--- SysQ	0.418	0.060	60.481	***	Supported
PEU <--- SerQ	0.122	0.042	20.721	0.007	Supported
PU <--- PEU	0.329	0.053	60.234	***	Supported
PU <--- GR	0.234	0.044	40.864	***	Supported
PU <--- SI	0.136	0.035	30.410	***	Supported
PU <--- InfQ	0.265	0.057	40.401	***	Supported
PU <--- SysQ	0.109	0.058	10.790	0.073	Supported
PU <--- SerQ	−0.054	0.036	−10.415	0.157	Rejected
PU <--- AW	0.176	0.037	40.199	***	Supported
In <--- PEU	0.082	0.057	10.449	0.147	Rejected
In <--- PU	0.716	0.068	100.404	***	Supported

Table 3 Korea structural paths assessment test

	Estimate	S.E.	C.R.	p	Results
PEU <--- InfQ	0.563	0.103	5.411	***	Supported
PEU <--- SysQ	0.207	0.087	2.082	0.037	Supported
PEU <--- SerQ	0.156	0.059	2.382	0.017	Supported
PU <--- PEU	0.2	0.083	2.423	0.015	Supported
PU <--- GR	0.377	0.083	4.024	***	Supported
PU <--- SI	0.168	0.048	3.281	0.001	Supported
PU <--- InfQ	0.418	0.112	3.703	***	Supported
PU <--- SysQ	−0.088	0.085	−0.91	0.363	Rejected
PU <--- SerQ	−0.01	0.056	−0.163	0.871	Rejected
PU <--- AW	0.122	0.056	2.191	0.028	Supported
In <--- PEU	0.213	0.078	2.444	0.015	Supported
In <--- PU	0.592	0.09	5.884	***	Supported

higher than 0.1 [2]. Collectivism/individualism cannot moderate the relationship between perceived ease of use to intention to use and perceived usefulness to intention to use E-government in the comparison between Korea and Vietnam because p-values are higher than 0.1 [2]. Masculinity/femininity cannot moderate the relationship between perceived ease of use and intention to use as well as perceived usefulness and intention to use E-government in the comparison between Korea and Vietnam because with p-values higher than 0.1 [2]. Long term orientation/short term orientation cannot moderate the relationship between perceived ease of use and intention to use as well as perceived usefulness and intention to use E-government in the comparison between Korea and Vietnam because p-values are higher than 0.1 [2].

Table 4 Vietnam structural paths assessment test

	Estimate	S.E.	C.R.	p	Results
PEU <--- InfQ	0.131	0.064	1.69	0.091	Supported
PEU <--- SysQ	0.547	0.083	6.187	***	Supported
PEU <--- SerQ	0.139	0.056	2.114	0.035	Supported
PU <--- PEU	0.322	0.066	4.891	***	Supported
PU <--- GR	−0.058	0.072	−0.848	0.396	Rejected
PU <--- SI	0.177	0.051	2.695	0.007	Supported
PU <--- InfQ	0.146	0.063	1.914	0.056	Supported
PU <--- SysQ	0.39	0.078	4.693	***	Supported
PU <--- SerQ	0.13	0.048	2.315	0.021	Supported
PU <--- AW	0.088	0.081	1.281	0.2	Rejected
In <--- PEU	0.154	0.079	2.053	0.04	Supported
In <--- PU	0.585	0.091	6.73	***	Supported

Indulgence/restraint cannot moderate the relationship between perceived usefulness and intention to use E-government in the comparison between Korea and Vietnam because p-value is higher than 0.1. However, indulgence can be the moderator variable of the relationship between perceived ease of use and intention to use in the comparison between Korea and Vietnam because p-value is 0.017 (<0.1) [2].

Table 6 shows that power distance does not affect the relationship between perceived usefulness and intention to use E-government for Korean (p-value = 0.202). Diversely, when p-value is less than 0.1, the power distance has negative effect on the relationship between perceived usefulness and intention to use E-government for Vietnamese (Beta = −0.133, p-value = 0.074) [2]. Indulgence/restraint has positive effect on the relationship between perceived ease of use and intention to use E-government for Korean (Beta = 0.045, p-value = 0.085). Diversely, when p-value is more than 0.1, indulgence/restraint does not have effect on the relationship between perceived ease of use and intention to use E-government for Vietnamese (p-value = 0.864) [2].

5 Conclusion

Based on the result, in case of the main model, the perceived usefulness factor is the strongest factor in both in Vietnam and Korea (0.716). This indicates that citizens will use E-government system/services when they get the benefits. After that, the perceived ease of use factor also has strong effect on the perceived usefulness factor. It means that when citizens feel any ICT system is easy to use, they will get their usefulness. Besides, the quality group with three factors which are services quality, information quality, and system quality also have strong effect to the perceived ease

Table 5 Checking moderator factor

Power distance

Model	NPAR	CMIN	DF	p	CMIN/DF	
	255	3893.102	1956	0.000	1.990	
		Estimate	S.E.	C.R.	p	Results
In <--- PEUxPD		0.031	0.024	1.312	0.190	Rejected
In <--- PUxPD		−0.109	0.031	−3.466	***	Supported

Uncertainty avoidance

Model	NPAR	CMIN	DF	p	CMIN/DF	
	255	4587.417	1956	0.000	2.345	
		Estimate	S.E.	C.R.	p	Result
In <--- PEUxUA		0.029	0.035	0.851	0.395	Rejected
In <--- PUxUA		−0.016	0.032	−0.509	0.611	Rejected

Collectivism/individualism

Model	NPAR	CMIN	DF	p	CMIN/DF	
	255	4444.614	1956	0.000	2.274	
		Estimate	S.E.	C.R.	p	
In <--- PEUxCO		0.035	0.032	1.114	0.265	Rejected
In <--- PUxCO		0.044	0.027	1.621	0.105	Rejected

Masculinity/femininity

Model	NPAR	CMIN	DF	p	CMIN/DF	
	255	4027.732	1956	0.000	2.059	
		Estimate	S.E.	C.R.	p	Result
In <--- PEUxMA		0.021	0.024	0.890	0.374	Rejected
In <--- PUxMA		−0.065	0.041	−1.578	0.115	Rejected

Long term orientation/short term orientation

Model	NPAR	CMIN	DF	p	CMIN/DF	
	255	4350.724	1956	0.000	2.224	
		Estimate	S.E.	C.R.	p	Result
In <--- PEUxLT		0.039	0.031	1.239	0.215	Rejected
In <--- PUxLT		0.036	0.027	1.337	0.181	Rejected

Indulgence/restraint

Model	NPAR	CMIN	DF	p	CMIN/DF	
	255	4046.794	1956	0.000	2.069	
		Estimate	S.E.	C.R.	p	Result
In <--- PEUxID		0.059	0.025	2.381	0.017	Supported
In <--- PUxID		−0.027	0.025	−1.084	0.278	Rejected

Table 6 Checking national culture dimensions difference between Korea and Vietnam

	Korea		Vietnam	
	Power distance	Indulgence/restraint	Power distance	Indulgence/restraint
In <--- PU	00.555 (p = ***)		00.638 (p = ***)	
In <--- PD	00.006 (p = 0.865)		00.006 (p = 0.912)	
In <--- PUxPD	−00.042 (p = 0.202)		−00.133 (p = 0.074)	
In <--- PEU		00.215 (p = 0.007)		00.282 (p = 0.001)
In <--- ID		00.018 (p = 0.588)		−00.038 (p = 0.645)
In <--- PEUxID		00.045 (p = 0.085)		−00.010 (p = 0.864)

of use with high value which are 0.122, 0.307, 0.418 respectively. Comparing the factors in quality group, system quality is the strongest which indicates that the role of the system is continuous and safe [2]. On the other hand, information quality also has the important role which affect both Perceived Usefulness and Perceived Ease of Use with the value of 0.265 and 0.307 respectively. In this research, in Environment group, the government regulation is the strongest factor which has direct affect to the perceived usefulness. Regulation is very important for the government because with the good regulation, government can supervise their state officials and their citizens. Besides, awareness and influence statement have the similar value with 0.176 and 0.136. For the E-government, awareness factor explains how people can understand the ICT services. Similarly, social influence statement strongly affects the perceived usefulness. If some people do not know about online services, the introduction from other people can help those citizens [30]. Regarding the comparison between Vietnam and Korea, there are many differences. For environment group, in Vietnam, the government still does not have good regulations to implement the E-government services/system. People do not have good condition to apply and use the new services [2]. Besides, the research shows that people in Vietnam can get the benefit and information of the E-government services/system form the others because the social influence factor which is one of the important factors (p-value = 0.007).

Based on the National Culture Theory of Hofstede, which applies the main model, this research cannot find the moderation of long/short term orientation, uncertainty avoidance, and collectivism/individualism from the perceived usefulness and perceived ease of use to intention to use the E-government (p-value > 0.1). Moreover, indulgence/restraint only moderates the effect of perceived ease of use to intention to use and power distance only moderate the effect of perceived usefulness to intention to use E-government. Furthermore, In Vietnam, power distance factor only has negative effect to the relationship between perceived usefulness and intention to use. Meanwhile, it has no effect in Korea. The country with high power distance would have negative attitude for using and implementing E-government services [31]. Vietnam is the country with high power distance, people can affect the behavior intention to use in many ICT services [2]. The problem to implement the E-government in Vietnam also comes from their culture. Diversely, indulgence/restraint has positive effect to the interaction of perceived ease of use and intention to use E-government in Korea (p-value = 0.085). However, it has no effect in Vietnam. In the countries which have indulgence more than restraint, the ICT service quality is more relevant [32]. It is the reason why the public services and E-government services in Korea has the high ranking in the world.

References

1. Stoltzfus K (2005) Motivations for implementing E-government: an investigation of the global phenomenon. In: Proceedings of the 2005 national conference on digital government
2. Van HT, Kim B, Lee SY, Gim GY (2019) The difference of intention to use E-government based on national culture between Vietnam and Korea. In: 2019 20th IEEE/ACIS international conference on software engineering, artificial intelligence, networking and parallel/distributed computing (SNPD): 409–420
3. Yadav V (2011) Political parties, business groups, and corruption in developing countries. Oxford University Press, Oxford
4. Ferro E, Sorrentino M (2010) Can intermunicipal collaboration help the diffusion of E-government in peripheral areas? Evidence from Italy. Gov Inf Q 27(1):17–25
5. Lin F, Fofanah SS, Liang D (2011) Assessing citizen adoption of E-government initiatives in Gambia: a validation of the technology acceptance model in information systems success. Gov Inf Q 28(2):271–279
6. Muguti KP (2013) Mobile government adoption in Zimbabwe: lessons learnt from South Korea and South Africa. Global Information Telecommunications & Technology Program, School of Innovation
7. Schuppan T (2009) E-government in developing countries: experiences from sub-Saharan Africa. Gov Inf Q 26(1):118–127
8. Khuong MV, West DM (2005) E-government and business competitiveness: a policy review. VNCI Vietnam Competitiveness Initiative
9. Tuan ND (2007) E-government in Vietnam: an assessment of province websites (Master thesis). Gadjah Mada University, Indonesia
10. Delone WH, McLean ER (2003) The DeLone and McLean model of information systems success: a ten-year update. J Manag Inf Syst 19(4):9–30
11. Schaupp LC, Carter L, McBride ME (2010) E-file adoption: a study of US taxpayers' intentions. Comput Hum Behav 26(4):636–644
12. Floropoulos J, Spathis C, Halvatzis D, Tsipouridou M (2010) Measuring the success of the Greek taxation information system. Int J Inf Manag 30(1):47–56
13. Venkatesh V, Davis FD (2000) A theoretical extension of the technology acceptance model: four longitudinal field studies. Manag Sci 46(2):186–204
14. Taylor S, Todd PA (1995) Understanding information technology usage: a test of competing models. Inf Syst Res 6(2):144–176
15. Dourish P, Bellotti V (1992) Awareness and coordination in shared workspaces. In: Proceedings of the 1992 ACM conference on computer-supported cooperative work, pp 107–114
16. Hart P, Saunders C (1997) Power and trust: critical factors in the adoption and use of electronic data interchange. Organ Sci 8(1):23–42
17. Gefen D (2000) E-commerce: the role of familiarity and trust. Omega 28(6):725–737
18. Tan M, Teo TS (2000) Factors influencing the adoption of Internet banking. J Assoc Inf Syst 1(1):5

19. Kumar V, Mukerji B, Butt I, Persaud A (2007) Factors for successful E-government adoption: a conceptual framework. Electron J E-gov 5(1)
20. Davis FD (1989) Perceived usefulness, perceived ease of use, and user acceptance of information technology. MIS Q 13:319–340
21. Hofstede G (1984) Culture's consequences: international differences in work-related values, vol 5. Sage, London
22. Islam N (2004) Sifarish, sycophants, power and collectivism: administrative culture in Pakistan. Int Rev Admin Sci 70(2):311–330
23. Geert H, Jan HG (1991) Cultures and organizations: software of the mind. McGrawHill, USA
24. McCoy S, Galletta DF, King WR (2005) Integrating national culture into IS research: the need for current individual level measures. Commun Assoc Inf Syst 15(1):12
25. Ford DP, Connelly CE, Meister DB (2003) Information systems research and Hofstede's culture's consequences: an uneasy and incomplete partnership. IEEE Trans Eng Manag 50(1):8–25
26. Hofstede GJ, Minkov M (2010) Cultures and organizations: software of the mind. Revised and expanded, 3rd edn. McGraw-Hill, New York
27. Erumban AA, DeJong SB (2006) Cross-country differences in ICT adoption: a consequence of culture? J World Bus 41(4):302–314
28. Srite M (1999) The influence of national culture on the acceptance and use of information technologies: an empirical study. In: AMCIS 1999 proceedings, vol 355
29. Bouaziz F (2008) Public administration presence on the web: a cultural explanation. Electron J E-gov 6(1)
30. Al-Somali SA, Gholami R, Clegg B (2009) An investigation into the acceptance of online banking in Saudi Arabia. Technovation 29(2):130–141
31. Aida A, Majdi M (2014) National culture and E-government services adoption Tunisian case. Int J Bus Econ Strategy (IJBES) 1:4
32. Gracia DB, Ariño LVC, Blasco MG (2015) The effect of culture in forming e-loyalty intentions: a cross-cultural analysis between Argentina and Spain. BRQ Bus Res Q 18(4):275–292

Study on Business Strategy Quantification Using Topic Modeling and Word Embedding: Focusing on 'Virtual Reality' and 'Augmented Reality'

Siyoung Lee, Sungwoong Seo, Hyunjae Yoo, and Gwangyong Gim

Abstract This study analyzed articles related to "Virtual Reality" and "Augmented Reality," the main elements of the Fourth Industrial Revolution, and reviewed the views that the media has by using text mining techniques. For this, 14,443 news articles on "Virtual Reality" and "Augment Reality" from 2011 to 2020 were collected from the BigKinds of Korea Press Foundation. For the analysis of collected datasets, the LDA algorithm-based topical modeling techniques implemented in Python3 languages were used. The number of topics suitable for the collected dataset using coherence was determined to be 14. The calculated topic was given the title of the topic through expert survey, and word vector was created using Word2vec, one of the word-embedding techniques. This word vector was then used to quantify the analyzed topic. Through this, quantitative expressions of strategic perspectives were made and changes in topics around the year 2016 were visualized when the issue of the "fourth industrial revolution" began to be discussed.

Keywords Latent Dirichlet allocation · Topic modeling · Word2vec · Topic quantification

S. Lee · H. Yoo
Department of IT Policy and Management, Graduate School, Soongsil University, Seoul, Korea
e-mail: sean.lee.kr@gmail.com

H. Yoo
e-mail: callnet0@daum.net

S. Seo
Hanwha Systems, Seoul 05451, Korea
e-mail: swoongee@gmail.com

G. Gim (✉)
Department of Business Administration, Soongsil University, Seoul, Korea
e-mail: gygim@ssu.ac.kr

© The Author(s), under exclusive license to Springer Nature Switzerland AG 2021
J. Kim and R. Lee (eds.), *Data Science and Digital Transformation in the Fourth Industrial Revolution*, Studies in Computational Intelligence 929,
https://doi.org/10.1007/978-3-030-64769-8_8

1 Introduction

The term "Fourth Industrial Revolution" refers to the next-generation industrial revolution led by Artificial Intelligence, the Internet of Things, robot technology, drones, self-driving cars and Virtual Reality [1]. The term was first used by Klaus Schwab, the chairman of the Davos Forum held in Switzerland in June 2016. There are many technical components of the "fourth industrial revolution," and among them, the areas related to digital content are "Virtual Reality" and "Augmented Reality." When a new concept or technology is first introduced, the media or researchers produce numerous positive or negative information related to it. However, the actual assessment or measurement of that information is discussed qualitatively again only when the issue is raised many years later. It is only after the impact on technological and social changes is re-examined with statistics on individual industrial areas or social changes that it is accepted into the everyday life of society. As such discussions on technical and social issues are not currently being measured in an immediate and quantitative format, specific strategic responses to technical and social discourse at this point are not easy. Moreover, there is a risk of consumingly spreading responses to social and technological changes. More scientific solutions to prevent this would be to define what will follow through quantitative measurements of social and technical discourse. For this reason, a strategic framework for social and technological discourse is urgently needed, and this study aims to observe how each topic has changed over time by automatically extracting and quantifying key topics from news articles before and after 2016, the first public discussion point of the Fourth Industrial Revolution on the subject of 'Virtual Reality' and 'Augmented Reality'.

2 Background

2.1 Business Strategy and Text Analysis

Recently, Business strategy is a business administration subject that many researchers have studied for a long time. It can be studied as a research topic in the realm of strategy management or business intelligence and used as a function of commercial software [2]. In the area of strategic management, there have been many cases in which business strategy frameworks have been studied using text analysis over the past decade. White et al. analyzed trends in the study of Information Strategy Management (ISM) from 2000 to 2013 using text analysis and bibliography [3]. Seol et al. used the patented information to analyze and identify new business areas using textual analysis techniques [4]. In addition, Lau et al. used the text analysis method to study the application of the hotel industry [5], Fan et al. analyzed the text analysis software and considered tools to help make strategic decisions [2]. A recent meaningful study was an analysis of energy industry areas using LDA, a technique of topical modeling. It used predefined and quantified keywords to analyze trends and status of enterprises or industries [6].

2.2 Text Analysis and Topic Modeling

In Many previous studies have been conducted on technology trend analysis using text analysis. Some studies have been done on text analysis itself, but some researchers, not computer-related researchers, have used specific techniques in text analysis to analyze topics, trends, or social likes or dislikes on specific topics. In that way, the main focus was simply to observe or extract trends in major keywords based on the frequency of the emergence of key keywords in each period [7] and to make visual representations through word cloud [8]. In some cases, the sentimental analysis based on expressions was done [9]. In Korea, text analysis techniques have been used as a way of analyzing research trends, especially in the areas of education [10], policy [11], and economy [12] as well as trends in technology [13]. Topic modeling, one of these methods of text analysis, is a statistical model for discovering abstract "topic" of a set of documents [14]. In other words, it is a technique that uses a statistical algorithm to clusters and analyzes them around topics that are key to an unstructured set of documents [14]. In Korea, research trends on AI (Artificial Intelligence) technology [13], research trends related to record management studies [15], and research trends on FinTech technology [16] were also analyzed using topical modeling. Topic modeling focuses on grouping documents with similar themes into each cluster in a large set of unknown documents, with related techniques such as LSI (Latent Semantic Indexing) and LDA (Latent Direchlet Allocation) [17]. In this study, LDA was applied as a topical modeling technique, which is a technique that deduces hidden topics from a set of documents using the probability of keyword appearance in the document. This technique is based on the premise that when a document is created, it is written with a keyword containing the subject.

2.3 Word Embedding

The traditional text analysis methodology used the frequency of the emergence of noun-oriented keywords extracted through pre-processing in the analysis. Examples of the tradition methodology include the Information Retrieval Model [18] that utilizes Term Frequency-Inverse Document Frequency (TF-IDF) as a Feature Value, and word cloud [8] that visually processes the frequency of keywords to provide an intuitive understanding of the nature of a group of documents. These techniques assumed keywords extracted through preprocessing as variables and conducted various analyses, but due to the existence of too many keywords in the nature of text analysis, there was a limit to treating them as variables. Many studies have been conducted to overcome this limitation, and in recent years, linear algebra and neural network methods have begun to emerge which does not make the keyword itself as variable [19]. For example, it is a technique that vectorizes individual keywords of the corpus given according to dimension N specified by the researcher into keywords with dimension N. Word2vec and FastText are well-known examples of such technique. These techniques are called word-embedding in the sense that in the process of producing human-specified N-dimensional vectors, the information in the context in which keywords exist is incorporated into the N-dimensional to be vectored.

The advantage of word embedding method over traditional frequency-based methods is that the efficiency of the operation is increased by first enabling a small level of vector operation. Secondly, as the result of vectorization, similar words between keywords can be detected in vector space as a method of calculating the similarity between keywords. In this study, Word2vec is used among these techniques. The similarity between each keyword of the topics obtained through the LDA and any keyword representing the 2-dimensional axis (which is subsequently referred to as the 'strategic axis') specified by the researcher was measured. After, it was applied to quantify the topics by marking the location of each topic in the 2-dimensional strategic axis space.

3   Research Methods

This study analyzes social trends related to Virtual Reality and Augmented Reality among digital content. In detail, it analyzes before and after the year 2016, when the 'fourth industrial revolution' began to receive attention as a popular theme. The analyzed data for service trend analysis used news data from 1 January 2011 to 7 June 2020. In using news data for analysis, the analysis was divided into two news data as of 2016, when the media's interest in the "fourth industrial revolution" began to be noted. The two data sets are news articles from 2011 to 2015 and news articles from 2016 to June 10, 2020, to see how social interest before and after that shifted as of 2016.

Topic modeling was used as a method of analysis to secure topics that span the entire duration of news articles, and LDA (Latent Direchlet Allocation) was used as a specific technique. In addition, the keywords of the topics derived as a result of the LDA were reviewed and the names of each topic were confirmed through the survey. The correlation between each topic was then analyzed through an Inter-topic Distance Map (IDM), which visually shows the distance over the two-dimensional coordinate space [20].

In previous studies, research trends or policy trends were analyzed through IDM analysis through the scale of each topic or the proportion of keywords contained in each topic. This study took a step further from the existing research and analyzed where each topic moved on the strategic axis by quantifying it on the strategic axis using the Word2vec technique and expressing it on the strategic axis in two dimensions. Figure 1 illustrates the experimental design of this study.

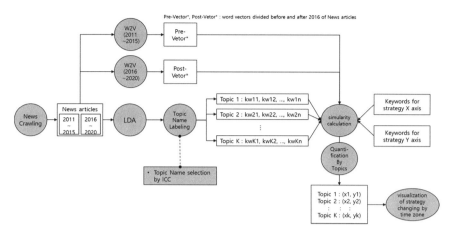

Fig. 1 Design of the experiment

3.1 Data Collection

To study the social trends of VR and AR, the study utilized news data from BigKinds, which is serviced by the Korea Press Foundation [21]. BigKinds is the nation's largest public news archive with the latest news from 54 major local media outlets, including newspapers and broadcasting. It is being updated daily and accumulating 60 million news stories for over 30 years. BigKinds is suitable for data analysis because it provides the original text of the news and a list of tokens extracted through form analysis for each news article. In this paper, "VR", "AR", "Virtual Reality", and "Augmented Reality" were set as search terms using the search function of BigKinds, and the period was set from 2011 to June 7, 2020, so that the trend of the four years before and after 2016 can be seen. In addition, given that the topics to be covered in this paper are digital contents such as VR and AR, but the base is on the field of technology, the press company was limited to Digital Times and Electronic Times only. The categories of news articles were limited to economics, society, culture, and IT science. As a result, 14,443 news articles were searched in 2011, 283 in 2012, 226 in 2013, 452 in 2014, 924 in 2015, 3017 in 2016, 2772 in 2018, 2908 in 2019, and 779 in 2020. The news data downloaded in the form of the Excel program consists of various columns for each news, such as date, media, classification, title, article text, keyword, and character extraction. In this paper, only the 'keyword' column was analyzed in consideration of the purpose of analysis.

3.2 Selecting Topic Title

Topics produced through the LDA are given a topic index expressed in numbers from 0 to 13 (the number of Topics K = 14). In the LDA technique, there was a study to automatically assign titles to each topic [22], but if the titles of the topics are expressed numerically, it is difficult for humans to intuitively understand or recognize them. To compensate for these shortcomings, this study conducted a survey on experts in the IT field to give each topic a title. The method of the survey was shown by listing 10 keywords containing 14 topics and respondents looking at a set of keywords were told to choose the appropriate title. Views presented as candidates for the title of the topic consisted of two to three title phrases, randomly combining keywords that formed the topic. The reliability verification that follows the title of the topic was verified using the Intraclass Correlation Coefficient [6, 23] and the reliability analysis function of SPSS 25.

3.3   Strategy Axis and Strategy Space

What sets this study apart from previous studies is that it defined the concepts of strategy axis and strategy space. The term 'strategic axis' in this study was defined as 'a linear set of similarity values between the strategic keyword and the topic keyword that exists within the word vector space'. Strategic space is defined as a 'two-dimensional plane represented by two strategic axes in a reciprocal relationship'. The key idea of this study is to measure the similarity between each of the more than one keyword related to strategy and the words that make up the LDA topics, calculate the mean, quantify and observe the change in the strategic space. For example, a bundle of words related to contrasting yet language-definable concepts, such as 'opportunity' and 'threat' and 'strong' and 'weakness' in SWOT analysis [24], can calculate similarities with other words within the word vector space [25]. At this time, the collection of similarity values calculated with the keyword 'opportunity', 'threat', 'strongness', and 'weakness' and the topical keywords in word vector space can be called strategic axes. For example, the two-dimensional coordinate space expressed by 'opportunity' and 'threat' can be defined as strategic space. These strategic axes inevitably differ in the composition of keywords depending on the environment that defines the strategic axis. If the strategic axis of 'opportunity' is assumed, the keyword that constitutes the strategic axis in the new product launch will be the word related to the actual expectations of the product, and the keyword that constitutes the strategic axis of the 'threat' will be the keyword related to regulation according to the market environment.

3.4   Word Embedding and Quantifying Topics

Word2vec was used as a word-embedding technique in this study. Word2vec is a method of learning and vectoring tokenized keywords in consideration of the order of appearance, as discussed in the consideration of prior studies. It is characterized by learning in the order in which the words appear, including information on the meaning of the sentences in which the words were included. Word2vec was used in two parts of this study. The first use was to select keywords to be used on the strategic axis, and the second was to quantify topics. The selection of keywords to be used on the strategic axis applied Word2vec for the entire duration of the news article, and then found words like the word 'public' and 'private' in word vector space and selected words that were actually semantically similar. When quantifying the topic, each of the top 10 keywords that make up each topic and each keyword that forms the strategic axis were measured in word vector space and the average of its similarity values was taken as the corresponding strategic axis of the topic. This is expressed in formula as follows:

$$\text{When Topic } K = \{w_1, w_2, \ldots, w_n\}$$
$$\text{Strategic Axis } S = \{s_1, s_2, \ldots, s_n\} S_x$$
$$S_x = \frac{\sum_{i,j=1}^{n} sim(wi, sj)}{i \times j}$$

It is expressed in such formula which allows each topic to be calculated as a corresponding value to one of the strategic axes.

Meanwhile, for this purpose of quantifying the topic, the news article data was divided into data from January 1, 2011 to December 31, 2015 and data from January 1, 2016 to June 7, 2020 respectively. Keywords on the strategic axis for each word vector corresponding to a divided period also measured the similarity in each word vector space with different periods. This is because they assumed that the news articles for the topics for that period might have different expressions. To make word vector, word vector was created by first making news data for the entire period and then dividing news articles into two data as of 2016. Both vectors, which are word vectors for the entire period and when they are divided, are set to be the same.

4   Experiment Result

4.1   Topic Analysis

The results of the topic modeling are shown in Table 1 to extract the central topics and related key words in news articles related to "VR" and "AR". Each topic was presented in order of proportion of topics using LDA-based topical modeling algorithms. The top 10 words by frequency, representing each topic, were extracted. A random title was then given around the association of the top 10 words by topic, and a survey was conducted on experts to verify the objectivity of the title. The survey of 22 experts surveyed 14 topics in the form of viewing keywords contained in each topic and selecting the most appropriate title. The Intraclass Correlation Coefficient (ICC) for

Table 1 Result of topic modeling on news articles

Topic	Topic name	Keyword
1	Government policy	Industry, government, ICT, sector, revolution, convergence, policy, drive, SW, innovation
2	Smartphone camera	Smartphone, product, camera, Samsung Electronics, launch, device, Google, use, Apple, video
3	Startup	Company, support, business, startup, start-up, investment, representative, www, center, global
4	4th industrial revolution	Ai, robots, services, iot, enterprise, platform, data, cloud, future, digital
5	Global investment	Market, business, investment, dollar, China, growth, sales, patents, displays, prospects
6	Smart information solution	Smart, information, solutions, utilization, application, deployment, management, automotive, possible, real-time
7	Mobile advertising service	Services, mobile, delivery, customer, information, smartphone, advertising, products, use, SNS
8	5G service	Service, 5g, network, standard, world, Olympics, speed, movement, commercial
9	Education program	Education, reality, students, universities, learning, programs, utilization, virtual, teaching, science
10	Video digital platform	Content, production, culture, media, video, digital, platform, business, realistic
11	Experience exhibition	Experience, exhibition, space, visitors, preparation, exhibition, booth, world, operation, participation
12	Mobile game market	Game, mobile, market, launch, representative, online, Chinese, character, pc, user
13	Carrier broadcasting platform service	Service, KT, SK Telecom, TV, provision, broadcasting, video, LGU+, content, LG
14	Event information	Event, hosting, hosting, theme, Seoul, introduction, participation, forum, industry, com

the survey results was relatively high at 0.75, confirming the reliability of the topic title selection through the survey.

Some of the calculated topical keywords such as com and www. seem meaningless. Presumably, keywords extracted from the website's address were not specifically excluded because they were not thought to have a significant impact on the topic assignment of topics.

4.2   Change of Topic in Strategic Space

Next, the paper will examine the change in location of topics by timing in strategic space. To express the strategic axis, 'public' and 'private' were first set as representative keywords for each strategic axis. Similar words were searched in the word vector space where the representative keywords you set were intended for the entire time period. Among them, keywords similar to representative keywords were visually selected, and each added three keywords to create a set of strategic keywords. The set of each strategic keyword is as follows:

```
strategy_word_X=['public', 'public institution',
'public sector']
strategy_word_Y=['private', 'private institution',
'private sector']
```

Next, a set of strategic keywords was quantified by calculating the individual keywords contained in each topic. The values of the topics calculated for word vectors can be expressed in the strategic space, as shown in Fig. 2. However, to avoid the complexity of expression on the chart, only 10 topics out of 14 topics, which account for 80% of the total keywords, were expressed in the strategic space.

As shown in Fig. 2, each topic was quantified according to the similarities between 'public' and 'private' and expressed in the strategic space, for example, in the case of 'government policy' topics located at the top right, the similarities between 'public' and 'private'. This means that in all news articles since 2011, 'government policy' has a high similarity with both 'public' and 'private'. On the other hand, the "global investment" topic has relatively high similarities with "private" and much less similarities with "public." The limitation in reviewing the full-time data is that the overall representation is similar to the post-2016 aspect because it exceeds the previous volume of news articles related to the fourth industry since 2016.

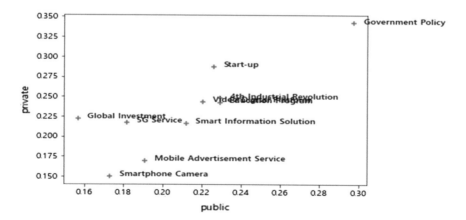

Fig. 2 Visualization of strategic space on the whole data

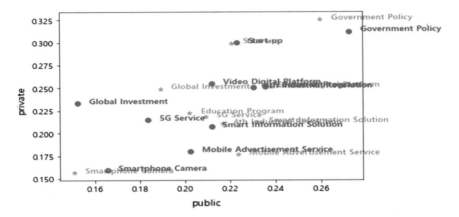

Fig. 3 Changes in topic in strategic space regarding 'public' versus 'private' before and after 2016

Next, quantified topics were expressed in strategic space by computing them with word vectors before and after 2016.

Figure 3 shows the location of topics before 2016 as '* (star)' and 'o' for topics after 2016. Overall, service and platform-related topics are relatively little different from "private" than before 2016, but the similarity with "public" has decreased significantly. In the case of the "fourth industrial revolution," the similarity between "private" and "public" radios has increased, which is estimated to be based on the public perception of the "fourth industrial revolution." The 'government policy' or 'start-up' topics have not changed much in the post-war period.

5   Conclusion

This study conducted data analysis focusing on news articles extracted by searching under the keywords "Virtual Reality" and "Augmented Reality." About 14,000 news articles from 2011 were analyzed through topic modeling, and various topics related to "Virtual Reality" and "Augmented Reality," which are central materials of this study, were widely discussed in news articles. In particular, it was confirmed that they were interested in the material base of digital contents such as telecommunication and video, and topics like utilization services. It was also confirmed that there was a topic on global investment due to the characteristics of digital content. Topics, the result of analysis using the LDA technique of topic modeling, were named through a survey, and reliability analysis was performed to ensure the reliability of the name. The named topics were easy ways to examine the change in location of topics in strategic space, as they were intuitively understandable to people. In addition, the collected news article data was word vectored using Word2vec, one of the word-embedding methods, and used to calculate the similarity between each keyword and topic for the strategic axis of 'public' and 'private' specified in this study. As a result, news article data showed that topics related to service platforms had become much distant from those related to 'public' compared to news articles before the 'fourth industrial revolution' was publicized. The topics on 'global investment' showed the same. However, the 'education program' topic and the 'start-up' topic was seen to form a more similar distance to the 'public' at least in detail. In particular, the 'education program' was seen to have a slightly higher similarity in terms of 'private'.

As shown in the research, this study is meaningful in quantifying and expressing the relationship between each topic, which is the result of the strategic axis and the topic modeling, in a two-dimensional space. It is expected that these techniques will serve as an aid to the determination of future strategic and policy-oriented decisions by the entity or government.

Future Challenges and Discussion Topics
This study showed that topics derived from topic modeling can be expressed in strategic space both nominally and visually. Nevertheless, the space on the strategic axis did not identify what factors caused each topic to shift at a point in time. This is a future research project that relates to explainable artificial intelligence (XAI), and requires a lot of research in the future. It is also necessary to apply the methodology of this study to many already proven business strategy frameworks. This is expected to serve as an opportunity to make wider use of this research's automated business strategy or text analysis as an aid to strategy establishment. These following topics should be discussed in future in connection with this study:

- Under what conditions and statistical methods will the homogeneity of the sample be demonstrated when there is a difference in the number of articles in time?
- How can scientific approaches and automation be realized in selecting keywords for the strategic axis?
- Can topics be named automatically?
- How do we verify the change in the location of topics in the strategic space?
- Can't we embed information other than the information obtained through the sentences in the news article in Word Vector?
- What should be considered when comparing entities such as enterprises or countries, rather than strategic pillars based on the concept?

In addition, the XAI and detailed research topics related to this study are expected to capture when and why future changes in corporate or government policies can be detected and responded to automatically in advance.

References

1. Schwab K (2017) The fourth industrial revolution. Currency
2. Fan W et al (2006) Tapping the power of text mining. Commun ACM 49(9):76–82
3. White GO et al (2016) Trends in international strategic management research from 2000 to 2013: text mining and bibliometric analyses. Manag Int Rev 56(1):35–65
4. Seol H, Lee S, Kim C (2011) Identifying new business areas using patent information: a DEA and text mining approach. Expert Syst Appl 38(4):2933–41
5. Lau K-N, Lee K-H, Ho Y (2005) Text mining for the hotel industry. Cornell Hotel Restaur Adm Q 46:344–362
6. Pröllochs N, Feuerriegel S (2020) Business analytics for strategic management: identifying and assessing corporate challenges via topic modeling. Inf Manag 57(1):103070
7. Hulth A et al (2001) Automatic keyword extraction using domain knowledge. In: International conference on intelligent text processing and computational linguistics. Springer, pp 472–482
8. Cui W et al (2010) Context preserving dynamic word cloud visualization. In: 2010 IEEE Pacific visualization symposium (PacificVis). IEEE, pp 121–128
9. Bakshi RK et al (2016) Opinion mining and sentiment analysis. In: 2016 3rd international conference on computing for sustainable global development (INDIACom). IEEE, pp 452–455
10. Kim YH et al (2019) Research subject trend analysis on educational innovation with network text analysis. J Educ Innov Res 29(1):91–116
11. Park JS, Hong SG (2016) Research of trends in Busan regional innovation policies using text mining. Korean J Local Gov Stud 20(1):1–20
12. Song H-J et al (2013) Trend analysis of Korean economy in the economic literature by text mining techniques. In: Proceedings of the Korean Society for Information Management conference. Korean Society for Information Management, pp 47–50
13. Chung MS, Lee JY (2018) Systemic analysis of research activities and trends related to artificial intelligence (aI) technology based on latent Dirichlet allocation (Lda) model. J Korea Ind Inf Syst Res 23(3):87–95
14. Steyvers M, Griffiths T (2007) Probabilistic topic models. In: Landauer TK, McNamara DS, Dennis S, Kintsch W (eds) Handbook of latent semantic analysis. Erlbaum, NJ. Information science in Korea using topic modeling. J Korean Soc Inf Manag 30(1):7–32
15. Park JH, Oh H-J (2017) Comparison of topic modeling methods for analyzing research trends of archives management in Korea: focused on LDA and HDP. J Korean Lib Inf Sci Soc 48(4):235–58
16. Kim TK, Choi HR, Lee HC (2016) A study on the research trends in Fintech using topic modeling. J Korea Acad Ind Coop Soc 17(11):670–81
17. Wikipedia, kr. Topic modeling. https://ko.wikipedia.org/wiki/%ED%86%A0%ED%94%BD_%EB%AA%A8%EB%8D%B8
18. Baeza-Yates R, Ribeiro-Neto B (1999) Modern information retrieval, vol 463. ACM Press, New York
19. Mikolov T et al (2013) Efficient estimation of word representations in vector space. arXiv preprint arXiv:1301.3781
20. Sievert C, Shirley K (2014) LDAvis: a method for visualizing and interpreting topics. In: Proceedings of the workshop on interactive language learning, visualization, and interfaces, pp 63–70
21. Korea Press Foundation, kr (2017) Big kinds. https://www.bigkinds.or.kr/
22. Lau JH et al (2011) Automatic labelling of topic models. In: Proceedings of the 49th annual meeting of the Association for Computational Linguistics: Human Language Technologies, pp 1536–1545

23. Bartko JJ (1966) The intraclass correlation coefficient as a measure of reliability. Psychol Rep 19(1):3–11
24. Hill T, Westbrook R (1997) SWOT analysis: it's time for a product recall. Long Range Plan 30(1):46–52
25. Shahmirzadi O, Lugowski A, Younge K (2019) Text similarity in vector space models: a comparative study. In: 2019 18th IEEE international conference on machine learning and applications (ICMLA). IEEE, pp 659–666

3D Printing Signboard Production Using 3D Modeling Design

Jungkyu Moon and Deawoo Park

Abstract As Seoul has become one of the world's top 10 cities, there are many demands for beautification and design for signboards. For business owners, the originality and publicity of the signboard are very important. In order to reflect these social demands, research using quaternary calculation technology for sign making is needed. In this paper, we design a 3D modeling signboard with slicing for 3D printing using a computer graphic. After the design is confirmed, modeling and printing for 3D printing are performed, and signboards are produced using acrylic, epoxy, and LED. LED signboards are divided into LED signboards for electric signboards and LED signboards for character channels. They display logo characters, etc., and research and manufacture signboards with high visibility and excellent advertising effects even at night. The signboard studied in this paper has the advantages of long life and low failure rate, and the disadvantage of high cost is to be solved by using 4th industrial technology.

Keywords 3D modeling · 3D printing · Fusion 360 · LED · Signboard · Slicing

1 Introduction

Signboards in the city's business district and commercial district are scattered, reducing the aesthetics of the city. As Seoul has become one of the world's top 10 cities, there are many demands for beautification and design for signboards. For business owners who use signboards, the originality and publicity of signboards is very important. For these social demands, research using 4th industrial technology [1] for sign making is needed. Fourth industries include 3D printers, drones, IoT

J. Moon · D. Park (✉)
Department of Convergence Engineering, Hoseo Graduate School of Venture, Seoul, Korea
e-mail: prof_pdw@naver.com

J. Moon
e-mail: prepel@naver.com

© The Author(s), under exclusive license to Springer Nature Switzerland AG 2021   109
J. Kim and R. Lee (eds.), *Data Science and Digital Transformation in the Fourth Industrial Revolution*, Studies in Computational Intelligence 929,
https://doi.org/10.1007/978-3-030-64769-8_9

(Internet of Things), smart farms, robots, information security, AI (Artificial Intelligence) [2], and big data. I will deal with 3D printers. Signage design and production using the latest industrialization technology is an industrial technology that can reflect the creativity of the entrepreneur and make the city's aesthetics beautiful.

In this paper, 3D modeling is designed using computer graphics. Therefore, the design that reflects the needs of the business owner can be changed. After the design is confirmed, modeling and printing for 3D printing are performed, and signboards are produced using acrylic, epoxy, and LED (Light Emitting Diode). There are various types such as printed flat signboard, LED signboard (Channel), neon signboard, post signboard, non-illuminated character signboard (SCSI), image monthly signboard, label signboard, window selection, and signboard. Among them, LED signboards are divided into LED signboards used for electric signboards and LED signboards used as letter channels. As they display logos and letters, the visibility is high even at night, so the advertising effect is excellent. The advantage is that the service life is long and the failure rate is low, but the disadvantage is that it is expensive. Fourth industrial technology is needed to further enhance the advantages and complement the disadvantages.

2 Related Studies

2.1 3D Modeling

Modeling is Model + ing, which means modeling, and literally means making a model when interpreted [3]. 3D Modeling refers to realizing 2D drawings in 3D. As software that implements 2D drawing, software such as Adobe's Illustrator and Photoshop can be representatively. 2D Drawings created through Illustrator, Photoshop, etc. can be saved as DXF (Drawing Exchange Format) files and used. In order to print with a 3D printer, 3D modeling of this DXF file is required. 3D Modeling Program Software is required for this task. Typical examples are Autodesk's Fusion360, Inventor, Auto CAD, 3D Max, and Software such as Maya.

In this paper, we will cover Fusion360. Fusion360 is a software that integrates CAD, CAM and CAE and is a Software capable of 3D Design & Modeling, Electronics, simulation, Generative Design, Documentation Collaboration, and Manufacturing [4]. Load the 2D Drawing worked on the Illustrator into the Fusion360 Software Work Space and proceed with 3D Modeling using 3D tools such as Extrude, Revolve, Sweep, Loft, etc. When modeling, modeling should be carried out in consideration of various factors such as material usage, printing time, and possibility of shrinkage during printing [5].

2.2   Slicing

Slicing means converting the 3D modeling file to G code so that the 3D printer can recognize and print the 3D modeling data. G code is a programming language used in numerical control, and is mainly used for computer aided manufacturing through automatic control machine tools.

In a fundamental aspect, G-code is a language that humans use to command how to make something on a computerized machine tool, and how to make it is defined as where the machine tool's tool moves and at what speed [6]. In a normal Slice Program, you can set the quality of production, amount of material, and printing time by setting layer height value, wall thickness value, infill density, printing temperature, flow, speed, travel, cooling, support, build plate adhesion type, etc. And Mold, Tree Support, Fuzzy skin, etc. Special functions can also be used.

The Slice Program includes 3D Slice Program software provided by each 3D Printer manufacturer.

In this paper, we will use Ultimaker Cura, a 3D printing slice software that is widely used worldwide. This program has recommended profiles obtained as a result of thousands of hours of testing, and about 400 settings can be set through 'Custom Mode', and the features and printing experience are improved through regular updates [7]. Cura basically has a Viewport that visualizes 3D Modeling Files of STL, OBJ, etc. You can adjust various options such as Layer Height, Line Width, Wall Thickness, Infill Density, and Printing Temperature. You can change it to Korean in the environment setting, so you can change it according to your convenience.

Clicking the Slice Button at the bottom right converts the 3D Modeling File in STL format into a G-code File that commands 3D Printer to print. In the G-code File, the temperature of the 3D Printer Nozzle is what, at what speed, and how much filament is extruded as show in Fig. 1. We can look at the expected movement path and stack structure through Preview, and depending on the production you want to print, you can set brim, raft, skirt, etc. on the bottom surface. Through this, the bottom surface of the production is not well adhered to the floor and is bent and floats, or the nozzle moves and touches the printing production. The case of losing can be compensated.

In addition, when you press the Slice Button, the time for the production to be printed and the required amount of filament are displayed at the bottom right, so you can check the expected filament consumption and time as shown in Fig. 1. G-code files can be printed by directly connecting to a 3D printer, or files can be moved by putting them on a USB or SD card.

2.3   3D Printing

3D Printer [8] is a manufacturing technology that creates a three-dimensional object by sprinkling a continuous layer of material, also called additive manufacturing [9]. For 3D Printing, 3D Modeling File, 3D Printer, and 3D Printer Materials in OBJ and

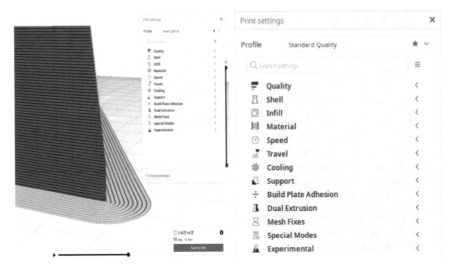

Fig. 1 Slicing in Ultimaker Cura

STL format are required, and a Slicing Software Program to send 3D Modeling File Data to 3D Printer is required. To do 3D printing, you first need a 3D printer. 3D Printer is a device that makes a three-dimensional structure by identifying products using CAD (Computer Aided Design) and printing materials layer by layer based on the data [10].

This is how 3D Printer prints production.

- FFF (Fused Filament Fabrication)
- FDM (Fused Deposition Modeling, Method)
- DLP (Digital Light Processing)
- SLA (Stereo Lithography Apparatus)
- SLS (Selective Laser Sintering)
- PBP (Powder Bed and inkjet head 3d Printing)
- MJM (Multi Jet Modeling)
- LOM (Laminated Object Manufacturing)
- Etc.

It varies according to the printing material and method, but in this paper, we will deal with 3D printer of FFF (FDM) method [11].

These are the materials used in 3D printers of the FFF (FDM) method.

- PLA (Poly Lactic Acid)
- ABS (Acrylonitrile Butadiene Styrene)
- NYLON, PVA, WOOD etc.

In this paper, we will deal with PLA (Poly Lactic Acid) [12].

Fig. 2 CAFÉ ADAGIO in PARIS BAGUETTE, ABC-MART DXF file imported into Fusion360

3   3D Printing Signboard Production 2D Design and 3D Modeling

3.1   3D Printing Signboard Production 2D Design

In 3D Printing Signboard Production, logos, characters, and fonts that individuals or companies want can be implemented relatively easily in shapes that are impossible in the existing industry, but designs that are difficult to apply due to complex processes even if possible.

In this paper, we intend to show the process of 3D Modeling, Slicing, and 3D Printing from 2D Design to PARIS BAGUETTE, ABC-MART, and MEDIHEAL, which are large franchises that are actually commercialized by using 3D Printer. 2D Drawing is necessary before 3D Modeling. People who need 3D Printing Signboard Production usually have their own logo or characters representing the company, and have their own logo and 2D DXF file of the character. Let's import this into the 3D Modeling Program Fusion360 Workspace as show in Fig. 2.

In the process of 3D modeling after importing a DXF file into Fusion360, an error may occasionally occur. After executing the Fusion360 Sketch mode on this DXF file, use the functions of Line, Spline, Project, etc. Do it as shown in Fig. 3.

3.2   3D Printing Signboard Production 3D Modeling

We are trying to 3D Model a newly sketched Drawing in Fusion360. For 3D Modeling, there are various functions such as Extrude, Revolve, Sweep, Loft, Rib, Web, etc. In order to use these functions, a profile is required. In Sketch, Line and Line or Line and Spline etc. Line and line meet to form a surface, and proceed with

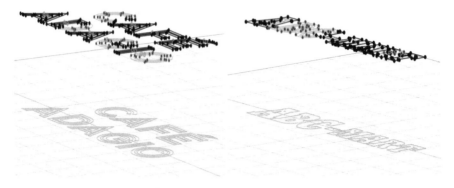

Fig. 3 A new sketch on the CAFÉ ADAGIO in PARIS BAGUETTE, ABC-MART dxf file imported into Fusion360

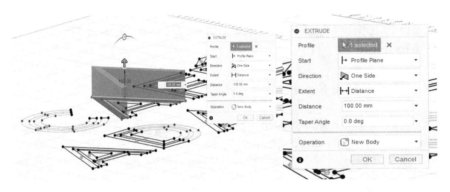

Fig. 4 Extrude drawing while holding a profile of CAFÉ ADAGIO in PARIS BAGUETTE

this surface. The Extrude function is to create a depth with the axis of Z in 2D drawing consisting of XY and 3D modeling as if it was a protruding shape as shown in Fig. 4.

Extrude can adjust the angle value in addition to the function of grabbing the profile and increasing it to the desired height as shown in Fig. 5.

The tools used and their values differ depending on the production method of 3D Printing Signboard Production desired by the company. For example, when you want to produce Acrylic 3D Printing Signboard, you need to proceed with 3D Modeling considering the part to apply Acrylic and where to put the LED, and if you want Epoxy 3D Printing Signboard Production, you need to understand the space and principle to pour Epoxy well as shown in Fig. 6.

Or, if there is a 3D Modeling File, you can load it as it is and proceed with the work as shown in Fig. 7. However, this may or may not be possible depending on the nature of the 3D Modeling Program, so you should use the 3D Modeling Program you need every time.

Fig. 5 Adjusting the angle value of the extrude function

Fig. 6 3D modeling of CAFÉ ADAGIO in PARIS BAGUETTE, ABC-MART drawing

Fig. 7 MEDIHEAL 3D
modeling imported 3D
modeling file into Fusion360

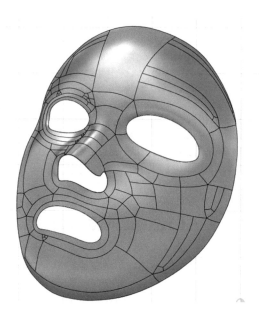

3.3 Slicing for 3D Printing Signboard Production Printing

When printing 3D Printing Signboard Production, the completeness depends on the Slicing setting value. In addition, it is important to reduce the quality and printing time in the printing of production. In general, if the printing speed is increased to reduce the printing time, the quality will be lowered.

To have good quality while reducing printing time, you can use various functions of Slicing. 3D Printing Signboard Production goes through a post-processing process, but let's unify the Layer Height to 0.2 mm. Sometimes the layer height is increased to reduce the printing time, but the quality of production is very poor and there is a high possibility of loss. Let the wall thickness be 1.2 mm. If the wall thickness is thin, the quality may be degraded because the infill line is reflected outside. Reduce the Infill Value to less than 10%. One of the advantages of 3D Printing Signboard Production is that it looks hard but light in weight. By reducing the infill value to less than 10%, the advantages of 3D printing can be utilized and printing time can be reduced. Since Printing Temperature differs according to the type of filament used, set the appropriate temperature. In general, I think you can increase the print speed to reduce the printing time. That is theoretically correct. However, increasing the print speed greatly reduces the quality of the production and increases the possibility of loss at the same time. Therefore, the print speed is set at 50 mm/s. Another important thing is Travel. When the nozzle travels, the filament can be retracted in the section where the filament should not come out, which prevents the formation of spider webs. Retraction can be set such as retraction distance, retraction speed, etc. This depends on the 3D printer and 3D printing situation.

The problem here is the combing mode. When executing combing mode, the nozzle moves over the printed area, avoiding the section that needs retraction. Then, the moving time becomes longer and the printing time becomes longer, but the quality of production can be improved by making less spider webs. Let the fan speed of Cooling be 100%. If cooling is not performed, the surface may flow down or the nozzle may be clogged, causing a risk of malfunction. Support depends on the case. If production can be printed without support, you can proceed without setting. Brim is recommended for the build plate adhesion type. This is a useful function to help settle the production. After completing the settings, press the Slice button to proceed with Slicing.

4 3D Printing Signboard Production Printing

4.1 3D Printing Signboard Production Printing

3D Modeling of 2D Drawing and sending Sliced G-code File to 3D Printer enables printing as shown in Fig. 8.

Fig. 8 MEDIHEAL being printed

4.2 3D Printing Signboard Production Post-processing Method

The production printed with the FDM type 3D printer is made of a laminated structure, so the laminated side is visible and the quality of the production is degraded. To supplement this, a post-processing method is required. In that way, various techniques such as a method using friction, a method using a chemical reaction, a method using putty and poly, etc. are used. As a post-processing method of 3D Printing Signboard Production, all technologies other than fumigation technology that utilize chemical reactions that are not suitable for filament of PLA components are used as shown in Fig. 9.

First, polish the laminated structure visible on the surface with relatively rough sandpaper (around 200 rooms), then apply a Surfacer to cover the surface, and polish again with medium rough sandpaper (around 400 rooms). It increases the roughness

Fig. 9 3D printing signboard production using sandpaper

Fig. 10 MEDIHEAL post-processing with putty method

of the sandpaper, and if this operation is repeated, the laminated structure becomes invisible.

If the size of 3D Printing Signboard Production required by companies is large, you can work with putty as shown in Fig. 10.

4.3 3D Printing Signboard Production, LED, Acrylic

When the post-processing method process of 3D Printing Signboard Production is finished, you can attach LED and Acrylic as shown in Fig. 11. The farther the distance between LED and Acrylic is, the more beautiful and spread out, but the height of 3D

Fig. 11 3D printing signboard production led, acrylic, epoxy applied

Printing Signboard Production increases, which is not good for aesthetics. Therefore, it is necessary to make the aesthetic as beautiful as possible by controlling the optimal height and light spreading effect.

4.4 3D Printing Signboard Production, LED, Acrylic

See Fig. 12.

Fig. 12 Constructed 3D printing signboard production

4.5 Comparative Analysis Table

The frame of the existing signboard is made of iron material, and the 3D Printing Signboard Production is made of PLA material, so the existing signboard is relatively heavy. Therefore, there are risks and labor costs are high in construction. The production unit price includes various expenses such as material cost, labor cost, utility bills, etc., but the frame of the existing signboard requires large equipment and a large space is required accordingly. In addition, labor costs are high because people work with the equipment themselves, and the work environment is noisy

Table 1 Comparative analysis table between existing signage and 3D printing signboard production

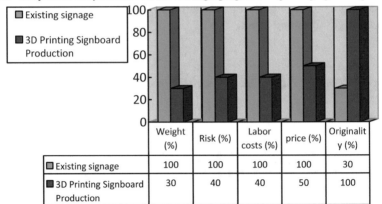

	Weight (%)	Risk (%)	Labor costs (%)	price (%)	Originality (%)
Existing signage	100	100	100	100	30
3D Printing Signboard Production	30	40	40	50	100

and messy. However, 3D Printing Signboard Production is produced by 3D Printer directly by pressing the Print button, and there is no need to watch it. In addition, it does not require a large space and the work environment is relatively quiet and clean.

Existing signboards can also be produced to give various shapes and effects. However, the resulting opportunity cost is large and there are forms that are impossible. However, in 3D Printing Signboard Production, if you consider the shape and effect you want and perform 3D Modeling, then the 3D Printer only needs to print as shown in Table 1.

5 Conclusion

Convergence with 3D Printer technology, one of the fourth industries, is very important to complement and highlight the weaknesses and advantages of the existing signboard production market. Through this, design that was previously impossible or difficult has become possible, and the weight is relatively reduced, thereby increasing safety and reducing labor costs. In addition, it is possible to achieve challenges that were difficult to try because it can significantly lower the cost of materials, production and construction. Through continuous research and development of various materials and designs, the 3D Printing Signboard Production industry in which 3D Printing technology is fused is expected to develop further. Furthermore, we will study 3D Printing Signboard Production that combines AI and IoT.

References

1. Younjung H (2017) A study on the artistic role of design according to the fourth industrial revolution. Korean Soc Sci Art 31:445–455
2. Suyoun C, Suyeon H, Deawoo P (2020) Platform design and source coding of AI responsive AR business cards. J Korea Inst Inf Commun Eng 24(4):489–493
3. Modeling. Namuwiki. https://namu.wiki/w/%EB%AA%A8%EB%8D%B8%EB%A7%81
4. Fusion 360. Autodesk. https://www.autodesk.com/products/fusion-360/overview
5. Sichang L, Gwangcheol P (2017) The ABS material characteristic compensation method for reflection 3D printing modeling compensation method for the design prototype production. J Integr Des Res 16(2):9–20
6. Jaewon L, Seokwoo S, Dosung C (2017) A study on slice program GUI design for FDM 3D print. J Integr Des Res 16(1):21–30
7. Cura. Ultimaker. https://ultimaker.com/ko/software/ultimaker-cura
8. Yonggoo K, Hyunkyu K, Geunsik S (2019) Strength variation with inter-layer fill factor of FDM 3D printer. J Korean Soc Manuf Process Eng 18(3):66–73
9. 3D printing. Wikipedia. https://ko.wikipedia.org/wiki/3%EC%B0%A8%EC%9B%90_%EC%9D%B8%EC%87%84
10. Youngmoon S (2015) Types and applications of 3D printers. Polym Sci Technol 26(5):404–409
11. Sukhan J, Juil Y (2017) Study of thermo plastic elastomer filament printing using 3D printing of FDM technology. Korean Soc Mech Eng 3299–3300
12. Jiae R, Sarang C, Jisoo P, Jihyo A, Jungmyoung L (2020) Changes in properties of 3D printing filaments by extruding at different temperatures and lignin contents. J Korea Tech Assoc Pulp Paper Ind 52(3):120–128

AI-Based 3D Food Printing Using Standard Composite Materials

Hyunju Yoo and Daewoo Park

Abstract 3D printing is one of the ways to advance the technology of the 4th indus-trial revolution. Instead of making a casting tool for the desired product, it directly produces the product through 3D printing. 3D printing can produce customized prod-ucts for each individual, so it is possible to construct a small smart factory. In partic-ular, AI (Artificial Intelligence) technology learns and judges legal judgments, cancer diagnosis, appropriateness judgments and standards for food ingredients, etc. that humans used to derive results. In the era of COVID-19, 3D food printing becomes an important turning point for non-face-to-face business and personalized business. 3D food printing is a technology that enables direct production of small quanti-ties using 3D digital design and personalized nutrition data. However, the current development stage of 3D food printing technology is only at the level of making a product with a simple form or only one material, and separate material processing is required to reach an appropriate level of print quality due to the printing charac-teristics of various food groups [1]. In addition, there are not enough structured data available for learning, and no prior development and indicators have been developed for standard composite materials that can be applied to various foods to reach print-ability. In this paper, we use AI machine learning to obtain adequate print quality in 3D food printing. We study supervised learning, unsupervised learning, and rein-forcement learning of AI machine learning, and design algorithms. In AI machine learning unsupervised learning, conformity and non-conformity are determined, and the result of the derived standard composite material value is applied to papers to evaluate printing adequacy. Through AI machine learning reinforcement learning, print aptitude is evaluated through rheological analysis, and big data values of various food groups applied with standard composite materials are secured.

Keywords Artificial intelligence · 3D food printing · Printability · Standard composite materials · Rheology · Hydrocolloids

H. Yoo · D. Park (✉)
Department of Convergence Engineering, Hoseo Graduate School of Venture, Seoul, Korea
e-mail: prof_pdw@naver.com

H. Yoo
e-mail: jar21@hanmail.net

© The Author(s), under exclusive license to Springer Nature Switzerland AG 2021
J. Kim and R. Lee (eds.), *Data Science and Digital Transformation in the Fourth Industrial Revolution*, Studies in Computational Intelligence 929,
https://doi.org/10.1007/978-3-030-64769-8_10

1 Introduction

Recently, the food industry is actively investing in biotechnology fields, such as applying ICT convergence technologies such as animal tissue culture and protein extraction, and food development applied from the mid-2010s. The market is projected to grow by 9.5% per year to $17.8 billion by 2025 [2]. The future key technologies of food manufacturing and processing can be summarized as nanotechnology, biotechnology, and 3D food printing.

However, the current development stage of 3D food printing technology is only at the level of making a product with a simple form or only one material, and separate material processing is required to reach an appropriate level of print quality due to the printing characteristics of various food groups [3]. In addition, there is not enough structured data that can be used for learning, and there are no prior developments and indicators for standard composite materials that can be applied to various foods to reach printability, so no machine learning algorithm has been developed accordingly.

In order to overcome these problems, the purpose of this study is to design two types of standard composite materials that can be applied to various food groups in order to obtain appropriate print quality in 3D printing, and classify them into 10 steps according to the concentration value, and a machine learning algorithm. Conformity and non-conformity are determined through the process, and the result of the derived standard composite material value is applied to papers to evaluate the print adequacy through unsupervised learning.

2 Related Studies

Until now, it has been pointed out that additive manufacturing processing technology in 3D food printing manufacturing is difficult to apply due to the fact that the food is composed of mixtures (carbohydrates, fats and proteins) and the inherent physicochemical characteristics of these ingredients [4]. The printability of a food matrix mainly depends on the rheological properties that are largely influenced by printing parameters [5]. Considering the consistency, viscosity, and coagulation properties of food ingredients, some food ingredients are stable enough to retain their shape after extrusion and lamination, such as chocolate, sugar, pasta, cheese, and mashed potatoes [6]. However, some foods such as rice, meat, fruits and vegetables cannot be printed easily, and food hydrocolloids and transglutaminases have been applied to many foods to improve extrusion and structural stability [7].

Soybean protein is a promising food ingredient in 3D food printing to improve print quality, but the relationship between food protein and printability is very limited. As a macromolecule essential for food structure, food hydrocolloids (polysaccharides, proteins or lipids) can be regarded as the skeleton of food structures, and have almost all processing, taste, nutrition and health benefits of food [8]. The future of food hydrocolloid research is worth looking forward to in the interaction between hydrocolloids and other food ingredients, the design of future functional food structures, and the regulation of interactions with the body [9]. 3D food printing can be applied to various food industries as it can produce individual foods with completely different tastes and flavors as well as ingredients of food. When applied to various food groups, the development of standard composite materials that can expect uniform print quality improvement is expected to be expanded to the general food industry, such as the formation of a new food culture, not the level of shaping food in three dimensions [10].

3 3D Food Printing System and Process

3.1 3D Food Printing System and Process

The 3D food printing system consists of a mechanical device, food material, and a program for 3D printing implementation. The mechanical device can be configured in various ways according to the shape and physical properties of the food material discharged by dispensing for dispensing, but it must be designed and manufactured through information on the viscosity and viscoelasticity that can control the dispensing speed and quantity. The discharging system is divided into melt discharging, liquid syringe discharging, and semi-solid extrusion discharging methods, and the range of solid is in powder form, liquid has a viscosity of 5–100cPs, and paste is suitable in the range of 500–40,000cPs.

In manufacturing food in 3D printers, selection of raw materials for food and information on properties of raw materials are important factors in order to make food raw materials printable. In addition, the pretreated food raw material must be stably maintained after being laminated in a plasticization or melting state while being supplied as a liquid or solid powder having flowable during the printing process. The shape of food can be maintained through reversible processing, printing temperature change, gelation and additives.

In this study, the 3D printing process was carried out using a self-developed extrusion-based 3D printer shown in Fig. 1, and the prepared standard composite material and food material are put into a syringe and moved to the nozzle tip to continuously extrude, fusing the previous layer, and designed. As a dispensing device for dispensing, a nozzle tip with a maximum volume of 60 ml and a diameter of 1.2 mm was used for 3D printing. All printing experiments were performed at room temperature, and slicing was controlled with open source software of CURA 15.04.6 (Ultimaker BV, Netherlands).

Fig. 1 Extruded FDM Type
3D food printer model used
in the experiment

3.2 Materials

3.2.1 Types of Food Additives for Standard Composite Materials Design

Unlike chocolate, which is extrudable and stable enough to extrude and maintain its shape after lamination, such as chocolate, some foods such as sugar, pasta, cheese, mashed potatoes and carbohydrates, fruits and vegetables cannot be printed easily, and the use of food additives is essential to improve extrusion and structural stability.

The soybean protein rich in essential and non-essential amino acids used in this study has excellent physicochemical and functional properties, and is a successfully printed material to form porous scaffolds.

In addition, hydrocolloids play an important role in the structure, processing, stability, flavor, nutrition, and health benefits of food, and thus are currently actively studied in the field of food science and technology. In food materials of 3D food printing, it is widely used for texture measurement by emulsification of liquid foods, stability of dispersion, thickening, gelling, etc., and has a property that tends to have a hydration layer by attracting water molecules around it because of its affinity with water. Raw protein, starch, gelatin, agar, and beet pectin are among the representative hydrocolloids.

In this study, standard composite materials include Soy Protein Isolate and hydrocolloid compounds, and hydrocolloids are divided into two groups. Gelatin (gelatin or gelatine) is a type of protein that has a transparent color and is mainly added to foods that give a chewy texture such as jelly because it has little taste and is decomposed by proteolytic enzymes (proteases). Alginic acid is a polysaccharide acid in the cell wall of brown algae. It is an acid contained in the cell walls of brown algae plants such as seaweed, seaweed, and kelp. The refined product is in the form of white powder. Alginic acid has a very wide range of uses. It is used as a paste when dyeing fabrics, is used to increase viscosity in ice cream, jam, mayonnaise, margarine, etc., and is also used in the production of lotions, creams, pills, and paper [11]. Carrageenan is a rubber-like substance collected from red algae plants and is used as a sticky material when making various foods such as chocolate, ice cream, syrup, and cheese.

3.2.2 Standard Composite Material Design for Improving Printability

The composite standard material (hereinafter SCM) was prepared as shown in Table 1, including Soy Protein Isolate and a hydrocolloid compound. It is divided into two types: isolated soy protein and hydrocolloid (Type A: gelatin+alginic acid/Type B: carrageenan+xanthan gum). The A-type sample is prepared so that sodium alginate is first dissolved in distilled water for 2 h with a stirrer so that the final concentration is 0.5%. Then, gelatin particles were added to the sodium alginate solution to reach a concentration of 1.0/2.0/4.0/6.0/10.0%, and the mixture was incubated in a water bath at 45 °C for 1 h. The following SPI powder was added to the sodium alginate and gelatin solution so that the final concentration reached 2.0/4.0/6.0/8.0/10.0%, and the sample names were SPI-GA1, GA2,…GA5. B-type is prepared so that the final concentration is 0.5/1.0/1.5/2.0/2.5% by dissolving the xanthan gum solution in distilled water with a stirrer for 2 h. Then, the carrageenan particles were added to the xanthan gum solution to reach a concentration of 1.0/2.0/4.0/6.0/10.0%, and

Table 1 Standard composite material and distilled water mixing concentration design

A-Type	Sample Name	Alginic acid	Gelatin	SPI
	SPI-GA1	0.5/99.5	1.0/99.0	2.0/98.0
	SPI-GA2	0.5/99.5	2.0/98.0	4.0/96.0
	SPI-GA3	0.5/99.5	4.0/96.0	6.0/94.0
	SPI-GA4	0.5/99.5	6.0/94.0	8.0/92.0
	SPI-GA5	0.5/99.5	10.0/90.0	10.0/90.0
B-Type	Sample Name	Xanthan Gum	Carrageenan	SPI
	SPI-CX1	0.5/99.5	1.0/99.0	2.0/98.0
	SPI-CX2	1.0/99.0	2.0/98.0	4.0/96.0
	SPI-CX3	1.5/98.5	4.0/96.0	6.0/94.0
	SPI-CX4	2.0/98.0	6.0/94.0	8.0/92.0
	SPI-CX5	2.5/97.5	10/90.0	10.0/90.0

Fig. 2 Composition of standard composite material samples (left: SPI-GA, right: SPI-CX)

the mixture was incubated in a water bath at 45 °C. for 1 h. Finally, SPI powder was added to the xanthan gum and carrageenan solution so that the final concentration reached 2.0/4.0/6.0/8.0/10.0%, and each sample name was SPI-CX1, CX2,...CX5 (Fig. 2).

3.2.3 Preparing Dough

In order to perform 3D printing by applying the standard composite material to the dough, the ingredient ratio was formulated as in Table 1 to determine the dough formulation, and to investigate the effect of the standard composite material on the printability of the dough. Were mixed at a mass fraction of 0.5, 1, 2, 3 based on the dough.

3.3 Rheological Properties

The rheological properties of standard composite materials and dough doughs of different compositions were analyzed using a rheometer (MCR 302, Anton Paar, Austria) equipped with a sandblasting parallel plate (PP25/s) with a diameter of 25 mm and a spacing of 1.

3.4 Statistical Analysis

For statistical analysis, analysis of variance was performed using the SPSS software package, and the significance of each sample was verified using the Duncan's multiple range test with $p < 0.05$ level.

Table 2 Dough formulas with standard composite materials (SCM)

Ingredients	Dough with SCM (g/100 g)										
	0	2	4	6	8	10	2	4	6	8	10
	0	SPI-GA1	SPI-GA2	SPI-GA3	SPI-GA4	SPI-GA5	SPI-CX1	SPI-CX2	SPI-CX3	SPI-CX4	SPI-CX5
Flour	45	43	41	39	37	35	43	41	39	37	35
Butter	20	20	20	20	20	20	20	20	20	20	20
Sugar	22	22	22	22	22	22	22	22	22	22	22
Milk	13	13	13	13	13	13	13	13	13	13	13
Total	100	100	100	100	100	100	100	100	100	100	100

4 AI-Based Printability Analysis of 3D Food Printing

4.1 AI Machine Learning Analysis and Algorithm Design

When designing an algorithm for AI machine learning, the size, quality, characteristics and available computation time of the data, and what you want to do with the data are the main determinants. As shown in Fig. 3, supervised learning of AI machine learning involves an input variable composed of previously classified training data and a desired output variable, and by analyzing the training data using an algorithm, a function that maps the input variable to the output variable can be found. In this paper, the algorithm of the supervised learning algorithm of AI machine learning uses a decision tree algorithm based on cases of mixing standard composite materials of 3D food printing, and the output value for the input value of 3D food printing. Perform pattern extraction to determine.

In this paper, supervised learning is conducted by classifying 3D food printing input value standard composite materials into 10 types, and the result is fitted by

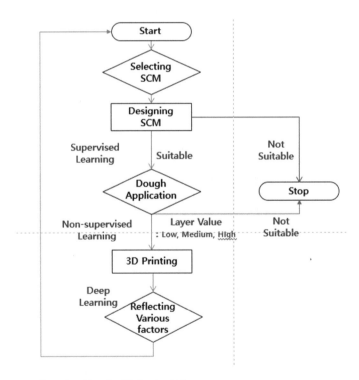

Fig. 3 Design of AI algorithm for 3D food printing

obtaining the difference between the predicted value and the actual value representing printing suitability and the square root of the mean through regression analysis. Divided by and nonconforming, values that do not indicate printability are excluded. The standard composite material determined to be suitable is applied to a dough prepared in advance to perform print aptitude evaluation, and the above print aptitude degree value is applied to evaluate it. The standard composite material function maps new cases from the training data and judges the results. In addition, to apply the unsupervised learning of AI machine learning, the 3D food printing aptitude is improved through data clustering and density estimation. As a result, the degree of 3D food printing aptitude is classified into Low, Medium, High, and Super High, and printing aptitude index is assigned. The unsupervised learning of AI machine learning extracts D food printing output pattern values from multi-class classification. The data of 3D food printing print aptitude is processed by standardization and normalization process appropriate for the label for AI reinforcement learning, and the print aptitude value is determined according to the change of type and density of standard composite material mapped by Big Data pattern analysis. Generate evaluation information (reward) that can be evaluated. It is possible to derive the printing aptitude and standardized big data of 3D food printing through repeated learning through regression by expanding the application of 3D food printing output information to chocolates, fruits and vegetables, and protein.

4.2 Analysis of Print Suitability of 3D Food Printing of AI Result Value

In order to find the optimum value of 3D food printing aptitude, AI supervised learning, unsupervised learning, and reinforcement learning methods were used. First, the measured values of shear modulus for the composite standard material classified into 10 indexes were measured as 593.54–4328.29cPs as shown in Table 3. A-type soybean protein, gelatin, and sodium alginate complexes exhibited higher shear modulus values than those of B-type soybean protein, carrageenan, and xanthan gum. The case that falls within the print suitability range of the viscoelastic modulus, which is directly connected to the hazardous shape retention force, is 500–40,000cPs, and the case that does not fall within the printability range is defined as Not Suitable. Therefore, among the composite materials, SPI-CX1 and SPI-CX2 are excluded from the print aptitude evaluation of standard composite materials. In addition, when included in a suitable evaluation index, unsupervised learning is applied to Low for 10,000cPs or less, Medium for 10,000–20,000cPs, High for 20,000–30,000cPs, and Super High for 30,000–40,000cPS. Such a process is standardized and applied even if it is not a standard composite material designed to improve the print aptitude used in this study, and is not suitable if it does not fall within the range of print aptitude through supervised learning. Because it is marked as, big data can be built.

Table 3 Evaluation of printability of standard composite materials

Classification	Sample name	SCM concentration (%)	Shear modulus (Pa)	Printability evaluation
A-Type	SPI-GA1	2	593.54	Low
	SPI-GA2	4	949.37	Low
	SPI-GA3	6	1974.29	Medium
	SPI-GA4	8	3354.97	Medium
	SPI-GA5	10	4328.29	Medium
B-Type	SPI-CX1	2	94.84	Not Suitable
	SPI-CX2	4	397.59	Not Suitable
	SPI-CX3	6	1002.54	Low
	SPI-CX4	8	2557.39	Medium
	SPI-CX5	10	3743.22	Medium

Table 4 Rheological properties and printability analysis of standard composite materials and dough mixtures

Classification	Sample name	SCM concentration (%)	Shear modulus (Pa)	Printability evaluation
A-Type	SPI-GA1	2	3584.98	Medium
	SPI-GA2	4	5854.33	Medium
	SPI-GA3	6	12652.96	High
	SPI-GA4	8	29772.33	High
	SPI-GA5	10	41382.58	Super high
B-Type	SPI-CX1	2	–	–
	SPI-CX2	4	–	–
	SPI-CX3	6	8492.99	Medium
	SPI-CX4	8	19321.58	High
	SPI-CX5	10	35326.22	High

Table 4 shows the result of measuring the shear modulus by applying the sample classified as suitable in the print aptitude evaluation of the standard composite material derived above to the dough. When various standard composite materials with print aptitude values were applied, the standard composite material value was added to the existing physical properties of the dough, resulting in a total elastic modulus of 3584.98–41,382.58cPs. When the same procedure as for the print aptitude evaluation of the above standard composite material was performed, SPI-GA5 was excluded because it deviated from the standard value of 41,382.58cPs. In addition, through the evaluation of unsupervised learning, print aptitude values were classified from Low to Super High.

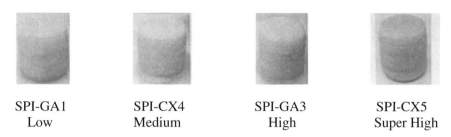

| SPI-GA1 | SPI-CX4 | SPI-GA3 | SPI-CX5 |
| Low | Medium | High | Super High |

Fig. 4 Dough printout with complex standard materials

The printout is a dough printout showing the print aptitude status from Low to Super High when the standard composite material is applied, and is a shape of a cylinder having a diameter of 3 cm and a height of 6 cm, respectively (Fig. 4).

4.3 AI-Based 3D Food Printing Prospect

The optimal value of 3D food printing aptitude was calculated as A/B type 10 sample values for AI supervised learning. The unsupervised learning of AI machine learning is classified as a multi-class classification, and it is classified into Low, Medium, High, and Super High to assign print aptitude indicators. 3D food printing reinforcement learning method was used to extract appropriate values for rheological properties and printability analysis of standard composite materials and dough mixtures.

The result based on the output of AI-based 3D Food Printing is shown in the figure. When the basic 3D Food Printing method and AI machine learning proposed in the paper are applied to supervised learning, unsupervised learning, and reinforcement learning. The advantages are as follows (Table 5).

5 Conclusion

In this study, in order to improve 3D food printing printing aptitude, standard composite materials composed of isolated soy protein and hydrocolloid materials were classified into two types, and 10 samples were made with different concentration values, and applied to dough to improve printability. This work is to overcome the limitation of 3D food printing, which cannot have standardized design values for printability due to the variety of food raw materials and materials added to improve printability. According to the existing 3D printing method, the print aptitude of the sample applied to the standard composite material and dough was measured, and according to the AI's supervised-unsupervised-reinforced learning design model, the appropriate-non-conforming, appropriate evaluation was performed. In future research, it is necessary to apply various hierarchical factors that affect the satisfaction of people's food provided by 3D food printing and food ingredients through AI deep learning reinforcement learning.

References

1. Kim C-T, Maeng J-S, Shin W-S, Shim I-C, Oh S-I, Jo Y-H, Kim J-H, Kim C-J (2016) Food 3D-printing technology and its application in the food industry. Korea Food Res Inst 2–10
2. Korea rural economic research institute: food technology status and challenges in the food industry-focusing on alternative livestock and 3D food printing. Rural Econ Res Inst Basic Res Rep 4–8 (2010)
3. Liu Y, Y Y, Liu C, Regensten JM, Liu X, Zhou P (2019) Rheological and mechanical behavior of milk protein composite gel for extrusion-based 3D food printing. Food Sci Technol 102:338–346
4. Derrossi A, Caporizzi R, Azzollini D, Severini C (2018) Application of 3D printing for customized food. A case on the development of a fruit-based snack for children. J. Food

Table 5 Comparison of advantages and disadvantages of conventional 3D printing method and AI application method

Comparison factor	Division	
	Conventional 3D printing method	AI supervised-unsupervised-reinforcement learning results applied
Printability evaluation	Conclusions are drawn through repeated experiments with a limited range of samples for various food ingredients and additives	When learning values that can improve various print aptitude such as food ingredient information, pretreatment status of food ingredients, additive information, temperature, printing speed, etc. through the process of guidance-unsupervised-enhancement, optimal information values can be obtained through vast amounts of big data
Application method of food additives	Various repeated experiments should be conducted to obtain limited information and quantified indicators obtained through a limited number of repeated experiments for the type and amount of additives designed by the experimenter	It is possible to input information values of various single and complex food additives that can improve printability, and obtain optimal printability index values according to the additive information through the above process
Predicting the implementation of the outcome	Material and physical conditions must be set for each individual situation in order to obtain optimal results	When developing UX/UI by applying and indexing data values for raw materials-food additives-physical condition values maximization of user convenience

Eng 65–75
5. Wang L, Zhang M, Bhandari B, Yang C (2018) Investigation on fish surimi gel as promising food material for 3D printing. J Food Eng 220:101–108
6. Schutyser MAI, Houlder S, de Wit M, Buijsse CAP, Alting AC (2018) Fused deposition modelling of sodium caseinate dispersions. J Food Eng 220:48–55
7. Guo Q, Ye A, Bellissimo N, Singh H, Rousseau D (2017) Modulating fat digestion through food structure design. Elsevier 68:109–118
8. Lua W, Nishinarib K, Matsukawac S, Fanga Y (2020) The future trends of food hydrocolloids. Elsevier 103:713–715
9. Ministry of science and technology information and communication: combination of advanced 3D printing technology and food tech! '3D Food Printer' (2020)
10. Chen J, Mu T, Goffin D (2019) Application of soy protein isolate and hydrocolloids based mixtures as promising food material in 3D food printing. J Food Eng 261:76–86
11. Fernandez C, Canet W, Alvarez D (2009) Quality of mashed potatoes: effect of adding blends of kappa-carrageenan and Xanthan gum. Eur Food Res Technol 229(2):205–222

Telemedicine AI App for Prediction of Pets Joint Diseases

Suyeon Han and Deawoo Park

Abstract Due to changes in lifestyles such as an increase in single-person house-holds and the spread of Corona 19, the number of families with companion animals is increasing. However, since companion animals cannot communicate with language, measures for diseases or pain that cannot be seen with the naked eye are inevitably insufficient. Among them, bone and joint disease affects the movement of companion animals, and better treatment is possible if the joint disease can be detected in advance in the stage before it worsens. In this paper, we intend to design a smartphone app based on 5G communication that can diagnose the presence and possibility of bone joint disease in companion animals. The Smarteck app takes photos of the back, side, and front of the dog on a smartphone connected via 5G mobile communication and uses the photographed photos and videos to have joint abnormalities such as patellar dislocation, hip dislocation, and cruciate ligament rupture. It will be designed to analyze and judge the presence of joint abnormalities through artificial intelligence (AI) machine learning. The Smarteck app is designed to identify joint disease abnormalities and possible conditions and provide joint disease information and treatment recommendations to dog guardian.

Keywords App · Artificial intelligence (AI) · Big data · Diagnosis · Smartphone · 5G mobile communication

1 Introduction

With the development of the 4th industrial revolution technology, the use of smartphones through 5G communication is becoming popular. In addition, 5G communication smartphones are advancing based on AI that executes AR VR through

S. Han · D. Park (✉)
Department of Convergence Engneering, Hoseo Graduate School of Venture, Seoul, Korea
e-mail: prof_pdw@naver.com

S. Han
e-mail: makee71@gmail.com

© The Author(s), under exclusive license to Springer Nature Switzerland AG 2021   137
J. Kim and R. Lee (eds.), *Data Science and Digital Transformation in the Fourth Industrial Revolution*, Studies in Computational Intelligence 929,
https://doi.org/10.1007/978-3-030-64769-8_11

high-speed video transmission and execution in real time and performs human tasks such as real-time map route search, mobile e-commerce, and interpretation. In particular, due to the COVID-19 PANDMIC incident in August 2020, there is a national support and social demand for non-face-to-face telemedicine using 5G communication smartphones.

As more and more people recognize companion animals as their families, interest in treating companion animals' diseases is also increasing. If non-face-to-face telemedicine can be used to treat pet diseases, it will be possible to block COVID-19 infection pathways that can occur face-to-face, thereby preventing COVID-19.

Since companion animals and guardians cannot communicate in human language, measures for diseases or pain that cannot be seen with the naked eye are inevitably insufficient. Among the diseases of companion animals, joint disease is a disease that affects the movement of companion animals. There may be a congenital joint abnormality, or it may be a disease caused by an acquired cause. If the caregiver can detect these joint diseases in advance, I think that the quality of life of the companion animal will improve and the happiness index of the caregivers will increase by selecting treatment at an appropriate time.

In this paper, we intend to design an AI smartphone app that can diagnose and predict the presence and possibility of joint disease in companion animals. The name of the app was named Smarteck by combining the English word "Smart" and "Check".

Smarteck App uses standards and data from professional veterinarians with more than 20 years of clinical experience. We will build big data and store it in the cloud.

After running the Smarteck app on your smartphone, take photos or videos of the back, side, and front of your dog, and match the photographed joint photos with normal joint photos to determine the presence of joint abnormalities such as patella dislocation, hip dislocation, and cruciate ligament rupture. do.

The Smarteck app is designed to determine the presence of joint abnormalities and joint diseases and the probability of diseases of the dog, and to recommend related treatments such as disease-related information, care method, and treatment timing to the companion animal.

2   Related Studies

2.1   Machine Learning

Machine learning, one of the artificial intelligence technologies, came into the 2000s and began to gain attention again. It is to change the parameters or structure of a system so that computers can perform the same or similar tasks more efficiently through experience and learning, and to create algorithms that use what they already know to infer what they don't know [1]. Machine learning methods can be broadly divided into (1) supervised learning, (2) unsupervised learning, (3) reinforcement learning, and (4) semi-supervised learning which is the middle between supervised learning and unsupervised learning.

Supervised learning is a method of injecting data composed of pairs of problems and answers into a computer to learn, so that when a similar new problem is given, the answer can be found. Unsupervised learning is a method of training computers to find answers by themselves and solve problems. Unsupervised learning includes clustering to classify data with similar characteristics and learning association rules to find relationships between characteristics in a large amount of data. Reinforcement Learning is likened to the process of learning about the world when a human baby is born. It is a method of finding a goal through the learning process, and a method of determining a policy so that the reward is maximized through actions and rewards accordingly in a specific environment.

In this paper, a dog joint abnormality diagnosis system app is designed using a data mining pattern recognition method to predict joint diseases in companion animals.

Figure 1 shows the components and processing process of the pattern recognition system. Pattern recognition systems are generally processed through five processes.

First, the object is measured in the real world, and the measured value is processed to some extent to make it into a desired shape or normalized through a preprocessing process that extracts a specific part.

After that, it is evaluated using an algorithm suitable for the type of problem, and to verify that the evaluation is successful, it is verified through a model selection process such as cross-validation or bootstrap, and results are derived [2].

Figure 1 shows the components and processing of the pattern recognition system. Pattern recognition systems are generally processed through five processes.

First, the object is measured in the real world, and the measured value is processed to a certain degree to make it into a desired shape or normalized through a preprocessing process that extracts a specific part. After that, it is evaluated using an algorithm suitable for the type of problem, and to verify that the evaluation is successful, it is verified through a model selection process such as cross-validation or bootstrap, and results are derived.

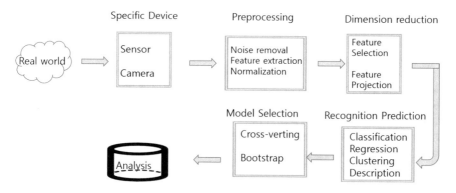

Fig. 1 Pattern recognition system components and processing

The Smarteck app in this paper is designed to be used in the steps before visiting a veterinary hospital by analyzing dog joint photos and videos to determine joint disease. It is necessary to build big data of dog joint photos and videos for analysis and determination of the disease.

2.2   Dog Joint Disease

Patella dislocation: refers to a disease that comes out of or out of the patella lying on the trochanter, and can be classified into 1st, 2nd, 3rd and 4th stages depending on the degree and pattern of the dislocation. In the first stage, dislocation occurs only when the patella is pushed inward or outward. Stage 2 can be intermittently dislocated even when the patella is bent and stretched. However, it refers to a state in which the patella returns to the normal position when the knee joint is moved to the normal position or the knee joint is moved again. In stage 3, the patella is mostly dislocated. Stage 4 means that the patella is permanently dislocated and cannot be moved to its normal position. Most of the causes of patella dislocation are related to musculoskeletal abnormalities such as medial quadriceps dislocation, the femur twisting and bending outward, the femur end abnormally formed, the rotation of the knee joint unstable, and the presence of deformities in the tibia [3].

Hip Dislocation: Hip dislocation is a disease that must be treated as soon as possible to prevent continuous damage to the soft tissue surrounding the hip and degeneration of the articular cartilage. When the hip joint is dislocated, it shows claudication in which weight support is impossible. When hip dislocation occurs, there is a difference in the length of the legs of both hind legs. The anterior dorsal displacement is shorter in the affected leg than in the normal leg and is reversed in the abdominal dislocation [4].

Cruciate ligament rupture: The cruciate ligament is the ligament that connects the femur and tibia and serves to limit the anterior and posterior movement of these two bones. It consists of an anterior cruciate ligament and a posterior cruciate ligament. Cruciate ligament injury means that the ligament is partially or completely torn, or the ligament that is attached to the bone is detached. If the anterior cruciate ligament is partially ruptured, a mild pain response or lameness may be observed. Partial rupture can proceed as a complete rupture over time [5].

Shoulder dislocation: It is not a common disease, but it refers to the separation of the humerus and scapula due to loss or damage to a part of the structure supporting the joint. It may be caused by trauma or may be congenital. Dislocation occurs when the biceps tendon and the forelimb ligaments above the inner and outer joints are torn or missing. He cannot support the weight of the dislocated leg and intermittently walks with his leg bent [6].

3 Dog Joint Disease AI Machine Learning Design

3.1 Big Data Analysis and AI System Design by Dog Joint Disease

In veterinary hospitals, treatment related to dog joint disease is generally conducted in the order of "interview with guardian → gait observation → pain area promotion → radiography → radiation reading → diagnosis with veterinary opinion".

In this paper, we use the pattern recognition method of data mining and design an AI app for dog joint abnormality diagnosis system so that companion animal joint disease can be predicted.

Table 1 shows the comparison between normal joint and medial and lateral patella dislocation or hip dislocation disease suspicious shape for data construction.

The probabilities in the left column of Table 1 can be classified by the point used when determining the presence or absence of a disease and the angle of the open joint of the point, and it is designed to be built into cloud big data through supervised learning of AI machine learning.

Points A and D denote the hip joint, B and E denote both knee joints, and C and F denote both ankle joints. In a normal dog joint arrangement (green line) without joint disease, the three joints above, hip joint, knee joint, and ankle joint lie on a nearly straight line. However, the joint arrangement (red line) of a dog suffering from joint disease goes out of line. The green line in Table 1 represents the normal arrangement, and the red line represents the joint arrangement of the diseased dog.

The angle represented by the line in Table 1 is represented by an equation in Table 2.

Table 1 Comparison of shapes for building big data by joint disease

Disease\Percentage	Medial patella dislocation (one side, two sides)	Lateral patella dislocation (one side, two sides)	Hip dislocation
50% ~ 90%			

Table 2 Parameters for analyzing the presence or absence of joint disease (angle measurement points)

$f(s) = X_L(\theta_{L1M}, \theta_{L3M}) + X_L(\theta_{R1M}, \theta_{R3M}) + X_L(\theta_{L1M}, \theta_{L3M}) + X_L(\theta_{R1M}, \theta_{R3M}) + X_L(\theta_{L1M}, \theta_{L3M}) + X_L(\theta_{R1M}, \theta_{R3M})$

θ_{L1M} = angle of red line AB and green line AB
θ_{L3M} = angle of red line BC and green line BC
θ_{R1M} = angle of red line DE and green line DE
θ_{R3M} = angle of red line DE and green line DE

3.2 Smarteck App Machine Learning Design

Design a smart app to be used in 5G communication smartphones. As a first step in Smarteck App, we design an AI machine learning method using a rule-based expert system. Figure 2 describes the five stages of a rule-based expert system. It can be designed in 5 steps as shown in Fig. 2.

As shown in Fig. 2, a rule-based expert system is built based on expert knowledge of veterinarians with more than 20 years of clinical practice. There may be some

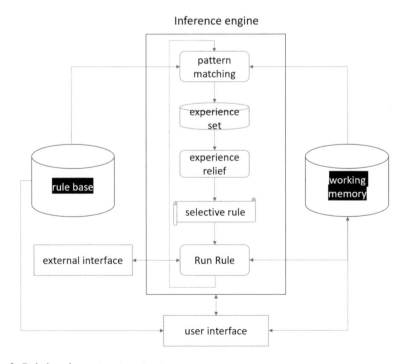

Fig. 2 Rule-based expert system structure

differences depending on the dog species, but in general, dogs consider the joint state in which the hip joint, knee joint, and ankle joint (ankle joint) are arranged in a straight line as a normal model. Figure 3 shows the normal joint position and arrangement. A and D represent the hip joint, B and E represent the knee joint, and C and F represent the ankle joint.

First, these three joint points are determined as a reference, and the position and arrangement are compared and analyzed to see how they differ from the normal appearance. For example, in dogs with a patella dislocation, the knee joint (points B, E) is pushed inward or outward, causing a deformation in the shape of the leg. In dogs with hip dysplasia or hip dislocation, the arrangement of the hip, knee and ankle joints deviated from a straight line when standing or walking due to a variation of the hip joints (points A, D).

Establish the data so far as a data base, and standardize the data format and contents by assigning a format to the data for each disease label. Based on the standardized data format and knowledge, it makes rules to train computers. Table 3 shows how the rule was applied to draw conclusions by implementing this rule as a rule to be used when judging unilateral patella dislocation and creating a chain of reasoning.

The big data on the patella dislocation and hip joint dislocation constructed in 3-1 is used by a computer to learn big data through machine learning. DB and machine learning compare and read the learned results, and access values of 50% or more and 99% or less through error judgment (false detection) and excessive judgment (overdetection).

The standardized data base is stored in the cloud through the Internet. The stored cloud information is subjected to AI machine learning supervised learning to generate

Fig. 3 Dog normal joint location and arrangement

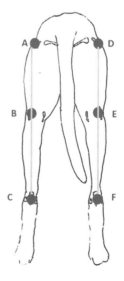

Table 3 Rules to use for smarteck App

IF	Both hip joints of the dog coincide with points A and D.
AND	The dog's left patella is located to the left of point B.
AND	The dog's left ankle joint is located to the right of point C.
AND	The dog's right patella coincides with point E.
AND	The dog's right ankle joint coincides with point F.
THEN	The dog may have a unilateral patella dislocation.

data for recognition and identification of joint diseases in dogs. Receives images of dog joint disease from the smartphone smart phone using 5G communication. Joint disease images are recognized by AI machine learning, identified, and transmitted to smart phone smart app.

Figure 4 is a flow chart of Smarteck App. Run the Smartekc app on your smartphone. Take photos or videos based on the shape of the dog's joints, and determine the presence or absence of a disease through the Smarteck app. When it is determined that there is a disease by removing error judgment and excessive judgment and selecting AI judgment, information and treatment related to disease and care are recommended.

3.3 5G Communication Design on Smarteck App

Figure 5 shows the communication design of Smarteck App.

The higher the quality of videos and photos, which are the data necessary for diagnosis, the more accurate it is to determine the presence or absence of a disease. To this end, it utilizes large-capacity, high-speed transmission of 5G communication, and improves the efficiency of image recognition and judgment on joint diseases with the big data of the cloud and the learning result of AI machine learning.

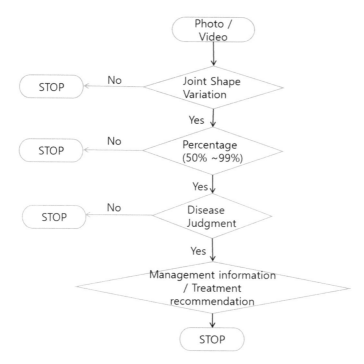

Fig. 4 Smarteck flowchart

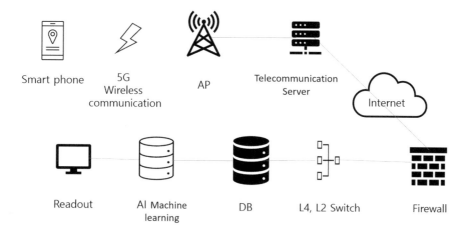

Fig. 5 Smarteck communication system

4   Design of Smarteck App for Determining the Presence of Joint Disease in Dogs

Install the Smarteck app on a 5G communication smartphone, sign up as a member, log in, and run the app. Figure 6 is the full screen configuration of the Smarteck app. After logging into the smart phone smartphone app, photograph the dog and scan the

4.1   Shooting Mode

joint shape. AI machine learning compares with the supervised learning data, reads whether the joint points match, and displays the read result on the screen.

As shown in Fig. 7, after running the Smarteck app on a 5G communication smartphone, access the camera app and take a picture or video of the dog. When taking pictures, take pictures so that both hind legs come out correctly. After shooting, upload photos and videos to the Smarteck app.

4.2   Scan Mode

The captured two hind legs are displayed on the screen and displayed by touching the app as shown in Fig. 8 based on the data of the hip joint, knee joint, and ankle joint of AI machine learning that are the standard for comparative analysis.

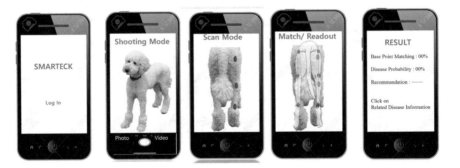

Fig. 6 Smarteck App

Fig. 7 Shooting mode

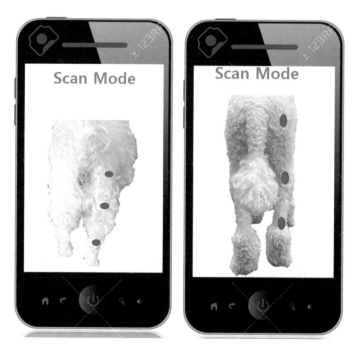

Fig. 8 Scan mode

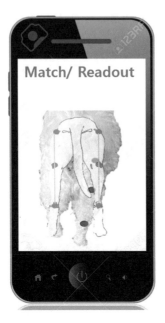

Fig. 9 Match/readout mode

4.3 AI Readout Mode

The information learned in AI machine learning is compared with the photographed joints. Compare and analyze the normal joint model and the point taken by touching in Fig. 8 as the pattern matching method.

Figure 9 is a screen under comparative analysis of dog joint disease.

4.4 Result Mode

The screen displays 4 items as the result of readout.

- What percentage does the reading match the shape of the normal joint?
- What percentage is the likelihood of having joint disease?
- What are your recommendations?
- Click for information on related joint diseases

Figure 10 is the result screen of the Smarteck app.

Fig. 10 Result mode

5 Conclusion

There is a social demand for non-face-to-face telemedicine with COVID-19 PANDMIC. In this paper, in order to solve the frustration of being unable to communicate about companion animal diseases through non-face-to-face telemedicine, we designed an AI machine learning method for predicting dog joint disease through the AI app Smarteck on a 5G communication smartphone.

Save the standardized data base in the cloud. Cloud information is based on AI machine learning supervised learning, which creates the base data for recognition and discrimination of joint diseases in dogs. Computers learn normalized and standardized big data for patella dislocation and hip dislocation of dogs using AI machine learning. Through 5G communication, images of dog joint diseases are transmitted from Smarteck App. Joint disease images are recognized by AI machine learning, identified, and transmitted to smart phone smart app. The learned results are compared and read by a computer using AI DB and machine learning supervised learning, and error judgment and excessive judgment are eliminated by AI deep learning, and the disease diagnosis probability is 50% or more and 99% value is approached.

Run the Smarteck app on your smartphone. A photo or video is taken based on the dog's joint shape, and the presence or absence of a disease is determined through the Smarteck app. It was designed to remove error judgment and excessive judgment, select AI judgment, and recommend disease and care-related information and treatment when it is judged that there is a disease.

While designing the Smarteck app, it was found that research data on the comparative angle of the normal joint position of a companion animal and the position of a diseased joint were insufficient. It is necessary to study future comparative angles by using AI deep learning technology for photos and videos that will be obtained through the Smarteck app.

References

1. Kun-Myung L (2019) Artificial intelligence-from turing test to deep learning saengneung publishing Co., Ltd. Gyeonggi-do
2. Bishop CM (2009) Pattern recognition and machine learning, Springer
3. Hamilton L, Farrell M, Mielke B, Solano M, Silva S, Calvo I (2020) The natural history of canine occult grade II medial patellar luxation: an observational study. JSAP 61(Issue4):241–246
4. Chun-ki C, Seok-joong K, Ji-yeon L, Jae-hoon L (2020) Dog home training, pp 97–99, Crown
5. Chun-ki C, Seok-joong K, Ji-yeon L, Jae-hoon L (2020) Dog home training, pp 93–94, Crown
6. Chun-ki C, Seok-joong K, Ji-yeon L, Jae-hoon L (2020) Dog home training, pp 103-104, Crown

Design of Artificial Intelligence for Smart Risk Pre-review System at the KC EMC

Youngjoo Oh and Deawoo Park

Abstract The world's major countries use Free Trade Agreements and protectionism to politicize non-tariff barriers along with tariffs that benefit their own industries. When it comes to world trade, there are a total of 79 legally mandatory certification systems in Korea. As part of the non-tariff barrier policy, electrical and electronic products are subject to pre-evaluation of the "Broadcasting and Communication Equipment Conformity Assessment" system. In this paper, we design Artificial Intelligence, Big Data, and Cloud Computing platforms to improve existing document evaluation. It analyzes big data in the form of electric and electronic products, and judges whether the pre-evaluation for the suitability of KC Electromagnetic waves intended to be sold in Korea is valid through AI machine learning. In addition, if it is not suitable, we intend to design an AI deep learning system that provides a complementary measure and provide it in the form of a Cloud Computing platform. We intend to develop a semi-automatic AI connection system that can perform KC conformity assessment for electric and electronic products to be sold in Korea.

Keywords Artificial intelligence · Big data · Cloud computing platforms · Conformity assessment · Free trade agreements · Harmonized system code

1 Introduction

This paper proposes a design of an AI that reviews the conformity assessment of a product in advance. In addition, the design model is explained by applying the laws on "KC Broadcasting and Communication Equipment Conformity Assessment" of Republic of Korea.

Y. Oh · D. Park (✉)
Department of Convergence Engineering, Hoseo Graduate School of Venture, Seoul, Korea
e-mail: prof_pdw@naver.com

Y. Oh
e-mail: oyj@skybrg.com

© The Author(s), under exclusive license to Springer Nature Switzerland AG 2021
J. Kim and R. Lee (eds.), *Data Science and Digital Transformation in the Fourth Industrial Revolution*, Studies in Computational Intelligence 929,
https://doi.org/10.1007/978-3-030-64769-8_12

151

Currently, there are 79 mandatory legal certification systems in 16 ministries and 107 voluntary statutory certification systems in 24 ministries in Republic of Korea. Even for the same purpose of "product safety" and "quality and performance assessment", there is an inconvenience of having to obtain duplicate certification because each department has different certification marks. As a result, not only time and money were wasted, but also international credibility was degraded, and local wealth leakage caused problems such as the need to re-certify because mutual authentication was not possible in cross-border transactions. Accordingly, 13 legally mandatory certification marks are integrated into one national integrated certification mark (2011.1). The initial national integrated certification mark (KC mark) was launched by combining 10 certification marks from the Ministry of Trade, Industry and Energy and the Ministry of Employment and Labor in September 2009. As of November 2011, 13 certification marks from 5 ministries were integrated into the KC mark. Has been. Since then, as each department established or changed the legal obligation certification system, the KC mark was introduced, and as of September 2017, 23 legal certification systems in 8 departments are using the KC mark.

In particular, the laws applied in this paper are operated under the "Radio Waves Act" and its subordinate laws as the "Broadcasting Communication Equipment Conformity Assessment System (Conformity Certification, Conformity Registration, Provisional Certification)" under the jurisdiction of the Ministry of Science and ICT, Technology and Communication. The products covered by this certification are electrical and electronic technical equipment and components, and some machinery. These products can be classified according to the Harmonized System Code. Perhaps, most likely, it will be included in the items from Part 16 to Part 20. However, as the market of IoT systems of the 4th industry expands, wireless communication functions are fused to other items, which are applicable to this law. The convergence of these products is making it difficult to determine the technical application and suitability of this Act. As such a solution, a "Smart Risk Pre-Review System" is used to judge suitability through preliminary review of the product. And check and supplement problems in advance. This will help you enter the market. This rule of review also applies to export products. As a result, it helps both domestic and foreign markets.

Because, the technical basis of the "KC Broadcasting and Communication Equipment Conformity Evaluation System" follows the conformity evaluation system of IEECC (IEC System for Conformity Assessment Schemes for Electrotechnical Equipment and Components) under the International IEC (International Electrotechnical Commission). Therefore, it may be easier to learn this AI further and expand its scope so that new products developed in Korea can be verified for the suitability of the international market.

Since the launch of the Trump administration in 2017, the United States has been around the world, including major allies. After the first 100 days, which caused numerous controversies, the Trump administration has attempted to change its foreign economic policy under the name of "America First!". The 'Pivot to Asia' and the 'American Pacific Century Initiative' were virtually ignored, and the Trans-Pacific Strategic Economic Partnership(TPP) withdrew from the Free Trade Agreement(FTA) for the first time in the history of US foreign policy without any congressional approval. Trump attacked immigrants from Mexico and promoted the United States-Mexico-Canada Agreement (USMCA) through amendment of the North American Free Trade Agreement. A 'trade war' was unfolded, involving simultaneous battles with several countries, including traditional allies in Asia and Asia.

The Trump administration abandoned its existing strategy to reconcile imbalances within a multilateral framework and adopted a bilateral strategy that puts pressure on individual states by focusing on the issue of the US balance of payments imbalance. The U.S. attacked traditional allies such as the European Union based on the unusual logic of placing a "national security threat" on trade issues, and a strategy of unilaterally imposing special tariffs on China without using the World Trade Organization based on Article 301 of the Trade Act of 1976. Took. The "trade war" between the United States and China lasted for more than two years, gradually having negative effects on the global economy, as well as a profound political and economic risk of the absence of a leading state in the world economy. In accordance with the US President Donald Trump's "America First" policy, the US government and China are in a trade war. This trade war is an economic war between the United States and China to preempt the economic advantage at the point of entering the stage of economic integration. As a result of the post-economic war, the world is expected to accelerate toward an integrated economic market. To take the initiative, a policy to reinforce tariff and non-tariff trade barriers that impose tariffs equivalent to 10–30% on each other is exchanged [1].

However, the trade war between the United States and China is causing a lot of damage to each other and adversely affecting neighboring countries. For this reason, this consuming trade war will soon end, and it is believed that economic integration will occur rapidly, centering on the interests and interests of the country. In particular, the US, which has chosen a bilateral strategy from a multilateral strategy, will achieve economic integration around the Free Trade Agreement, which represents bilateralism.

Now, we live in a different world than before, where new technologies such as 5G and cloud emerged in the era of the Fourth Industrial Revolution, where all people, not a few, simply and easily find and share information. As a result, each country on the planet has come to an era called the global information revolution. Powerful countries with such technological advantages have entered into a Free Trade Agreement that meets their own interests and interests, leading to protectionism. It is choosing a policy to lower the tariff barrier that used to protect its own industry. The Free Trade Agreement is the removal of various barriers to protection such as tariffs, which are trade barriers for free trade between the two countries. This has the advantage of allowing more free trade and exchange of products between countries, but there are many in Korea in that countries with weak industrial structures are concerned about collapse of the industrial sector and that countries with large amounts of capital and technology dominate the culture of the other country. It is controversial. Countries that have signed free trade agreements are building non-tariff barriers to protect their industries to avoid this situation. However, these non-tariff barriers dominate most of the market in large corporations that dominate capital and information and provide a reason for preventing SMEs from entering the market. The "Smart Risk Preliminary Review System" in this paper is a platform that provides and shares solutions for the Technical Barrier Trade (TBT). It is expected that this platform will perform the function of providing information democratization by sharing regulatory information of products.

Figure 1 is a flow chart of the "Smart Risk Preliminary Review System" applying the technical regulations and laws of the "KC Broadcasting and Communication Equipment Conformity Assessment System". It is designed to access and authenticate information of subscribers in the production, manufacturing, distribution, and customs fields of products using a block chain. Big data created in the "Smart Risk Preliminary Review System" is designed to be shared and operated in a cloud computing network. In the "Smart Risk Preliminary Review System", the HS code (Harmonized System Code; Harmonized Community Description and Coding System) is designed to use AI machine learning, and the search for technical

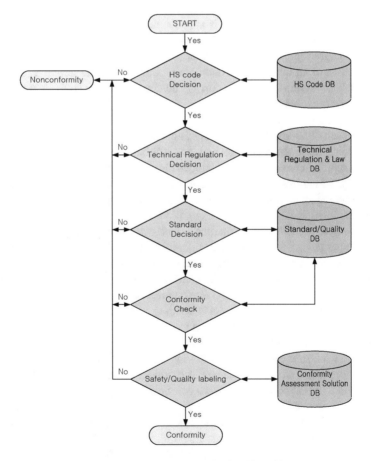

Fig. 1 The flowchart for smart risk pre-review system in the KC EMC

regulations and standards is designed to use AI deep learning algorithms. In addition, conformity verification AI machine learning can be learned, and conformance completion DB is designed to apply safety/quality labeling rules.

"Smart Risk Preliminary Review System" is a non-face-to-face conformance tester of products. This is because all AI engines are implemented with cloud computing. The AI design for the smart risk preliminary review system studied in this paper will be used as basic data for the development of public and private institutions related to trade.

2 Analysis of Smart Risk Preview System

To sell products or to sell imported products, technical regulations, standards, conformity assessment procedures, and safety labeling are defined in the importing country's legislation, as shown in Fig. 1. Must be checked for a conformity assessment. These product sales conformity assessment processes are defined in international standards, so that the Member State shall apply them in accordance with this standard.

There are compulsory implementations of technical regulatory requirements. In order to obtain the solution desired by the Smart Risk Pre-Review System, the technical information of the product manual, the parts list, the circuit diagram, the compounding cost, the manufacturing process diagram, the quality control documents, and the cyber security of the use system are required. These product preliminary reviews are digital documents written electronically that may be forged, tampered with, or compromised. In addition, as product preliminary review information is managed and distributed online, it is difficult to check the document history, thereby ensuring authenticity, integrity, reliability, and usability of digital documents. It must be necessary [2].

2.1 Digital Documents and Cybersecurity for Smart Risk Preview

In order to obtain the solution that users want to the" Smart Risk Pre-review System", technical information and manufacturing processes of products, such as instruction manual, component list, circuit diagram, and mix ratio, and cybersecurity for the system of users and security of quality control documents is required. This product pre-review information are digital documents written electronically, potentially tampering, and damaging. Furthermore, as product pre-review information is managed and distributed online, it is necessary to obtain authenticity, integrity, reliability, and usability for digital documents because document histories are difficult to verify.

2.1.1 Blockchain

Blockchain was first published by an anonymous developer named Satoshi Nakamoto in October 2008 in paper titled "Blockchain: A Peer-to-Peer Electronic Cash System [3].", in the cryptographic technological community "Gmane". The paper explained that double payments can be prevented by using P2P network link. Blockchain was selected as one of the 10 key technologies of the 4rth Industrial Revolution at the World Economic Forum in 2016. and they explained that the economic seize of the Blockchain will account for 10 percent of GDP [4]. As shown in Fig. 2. Blockchain consist of a header and body for each block. To prevent to falsify of block information, it includes electronic signature technology of participants in the encryption network of hash value data and in the encryption network key which is optimized for cyber-attack defense. In the Blockchain, hash values are converted hash values that are combined with the number of random specified lengths and characters to be used to determine falsification [5]. Blockchain can be used in various fields such as financial, economic, legal, electronic, voting public data management, etc. On the financial side, it is expected create new financial instruments by combine pitch and blockchain to P2P individual transaction method, not banking transaction method, activating cryptocurrency money, cryptocurrency exchange, distribution, automatic investment, etc.

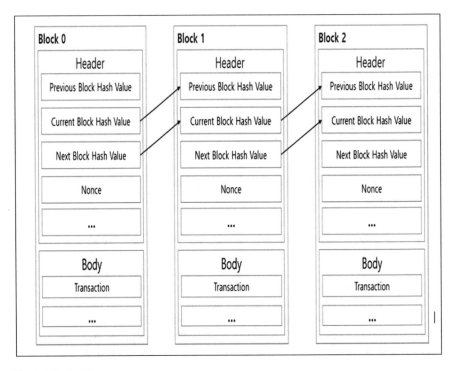

Fig. 2 Blockchain structure

As the central organization does not need function, the system will be newly constructed based on reliability using a P2P based on blockchain, which will have effect of maintenance cost and commission reduction in financial transactions. Also, a new digital market will be created, such as e-voting and public data management.

With the development of Artificial Intelligence and Cloud computing, The 4rth Industrial Revolution will bring about a major change in the social structure and actively promote the application of blockchain technology in the development of platform, device, application to prepare for the ripple effects of various industrial institutions in a centralized structure, thus leading the future ecosystem by developing new business creation and services, and is creating new revenue through continuous improvement [6]. However, there is a disadvantage that blockchain applies and storage cost increase as much as the number of users due to distributed storage of data, and the cost and the time of network processing are needed to send a computing power and a data to multiple locations simultaneously to store the same contents.

2.1.2 Blockchain Platform Classification

Blockchain which is currently being developed is built based on various platform for each location and is typically divided into Private and Public types. This is developed by utilizing the characteristics and scope of the blockchain participation network [7].

The Public Blockchain is a structure that anyone can allows participation in the network of who is responsible for the blockchain without the permission of the specific agency. The blockchain can be freely trade within the trade network. By using the consensus algorism such as PoW (Proof-of-work) and PoS (Proof-of-Stake), all users who participate in the public blockchain are shared all their trading books to ensure high transparency [8]. These features can have perfect security even for cyber-attacks, such as hacking, but slow network speeds, resulting in slow transaction speeds.

The Private Blockchain is a centrally blockchain unlike the public blockchain which does not share transaction details to all network users but provide high speed processing by closed companies and participation blocks. If the central server is under cyber-attack due to central control, all blockchain stored in server may not be recoverable the person responsible for cyber-attack has unclear shortcomings. The typical private blockchain is Hyperledger Fabric enveloped by Linux Foundation. Hyperledger Fabric is composed by the share ledger, personal information, consensus and smart contact. The composition of the share ledger is two share ledgers. This consists of a Transaction Log that records all transactions and the world state that stores a specific state [9].

Information is a comprehensive concept that includes information of various electronic types, such as images and voice, in addition to information of document types prepared by computer or information processing equipment. The overseas standard guidelines related to electronic document are 'IOS 15489' (Recorded Management Procedure and Method), which are international standards for records and management [10]. It was established under the supervision of 'SC 11' (Record

Management sub-committee) of 'IOS/TC46' (Information Technology committee). The posed Chapter is to protect all records and standardize the management process and procedure for more effective retrieval of evidence and information contained in records. 'IOS 15489' classifies the quality requirements of records management as "Progressive", "Reliability", " Integrity", " Availability" [11].

2.2 Analyze the Process of Product Sales Conformity Assessment

To sell a product or to sell an imported product, technical regulations, standards, conformity assessment procedures and labeling as specified in the laws of the importing country, such as Fig. 3. The product conformity assessment processes are defined by International Standards, so that Contracting States shall apply them in accordance with this standard in accordance with their own country.

The requirements for technical regulations are established by the Radio Wave Act, the system and the government led by regulations that describe product characteristics, related processes, and methods of production. The performance of the technical requirements is mandatory. If the requirements are not complied with, the company shall be fined, collect goods and be subject to legal restrictions.

Product characteristics, related processes and ISO standard production methods follow the standardization conditions. For the common and repetitive use of standards, certification bodies such as KTL, UL and TUV give approval. There is no coercion in the implementation of the standard. Producers and consumers gain reliability from the market by getting the standard certification.

The process of ensuring that the product meets the technical requirements of regulations and standards for the EMC is the conformity assessment. The conformity assessment process includes testing, testing, and certification.

The quality indication is to print the product's trademark, quality, content, etc. on the package or to mark it by affixing it. The reliability of the product is notified to the

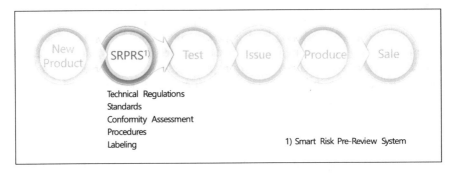

Fig. 3 The smart risk pre-review process

consumer by disclosing information about the suitability of the product's technical regulations and standard requirements to the consumer. Quality displays are subject to fines and mandatory warehousing because they set regulations in laws and technical standards.

3 AI Design of Smart Risk Pre-review System

3.1 The Machine Learning Design to Search HS Code

The Harmonized System Code (Harmonized System Code; Harmonized Community Description and Coding System) is a numerical code of the types of goods traded in foreign trade transactions, which was established by the International Convention in 1988. Basically, it consists of six digits, and each country is using four additional digits for detailed classification. Classifying all products by HS code is possible first, and all products worldwide are subject to tariffs in the basis of HS code. That is, the products traded in the world have HS codes.

The "Smart Risk Pre-review System" learns the rules by Machine-learning in each country's HS. Machine learning is designed to learn by the rules and to find HS codes of the products that are subject to.

The "Smart Risk Pre-review System" in this paper designs the Machine Learning API with the product information provided by the Smart Risk Preview System, HS code is found. To the Smart Risk Pre-review System to find HS codes, the Machine Learning API marked the documents to be learned in Fig. 4. Documents are the function and usage of the product, information of the target of use and the manufacturer's company, and information of the importer. Documents are classified and stored on a Data Base and used as Cloud Computing.

The product is defined by the product name and the product use manual in advance review application. In addition, the HS code of the manufacturer and importer shall be found in the importing country's position in accordance with the rules of HS code of the manufacturer and importer with the information of the manufacturer and importer. At the actual work site, customs officials are finding HS codes, feeling very difficult.

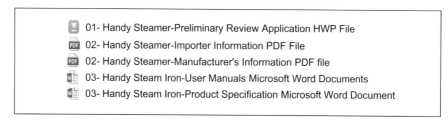

01- Handy Steamer-Preliminary Review Application HWP File
02- Handy Steamer-Importer Information PDF File
02- Handy Steamer-Manufacturer's Information PDF file
03- Handy Steam Iron-User Manuals Microsoft Word Documents
03- Handy Steam Iron-Product Specification Microsoft Word Document

Fig. 4 Example of an input document—HS code for smart risk pre-review system

In the era of super-connections of fourth industrial technologies, products become more complex, and customs agents who find HS codes are also designing machine-learning APIs to find HS codes in this paper to solve difficulties.

3.2 Deep Learning AI Design to Search for Technical Regulations and Standards

The product's HS code is determined by the AI engine designed in "Smart Risk Pre-review System". This system analyzes big data in cloud computing to find the correct code. The HS code for this paper sample is "8516.40-0000". Deep Learning AI design is designed to establish technical regulations and standards. Analyze big data from the data base stored in the cloud. Related laws are extracted from data mining from big data analysis. The contents of protecting the technical characteristics and regulations and standards of the product from HS code are designed with Deep Learning AI. Through HS code secured by Deep Learning AI, it is found that four related statutes may be involved as input values for electric iron "8516.40-000000".

- Electrical Appliances and Consumer Products Safety Control Act
- Radio Waves Act
- Act on the Resource Circulation of Electrical and Electronic Equipment and Vehicles
- Act on the Promotion of Saving and Recycling of Resources

However, the four technical provisions may or may not apply depending on the technical characteristics of the electric iron and the information of the importer. To establish technical regulatory requirements, the "Smart Risk Pre-review System" may search for and confirm related information in big data in the clouding computing.

First, it is designed to add circuit diagrams First, it is designed to add circuit diagrams, assembly drawings, and parts list as input information of the "Smart Risk Pre-review System" as Fig. 5.

01-Handy Steam Iron-Preliminary Review Application HWP File
02-Handy Steamer-Importer Information PDF File
02-Handy Steamer-Manufacturer PDF file
03-Handy Iron-User Manual Microsoft Word Document
03-Handy Irons-Specifications Microsoft Word Documents
04-Handy Steam Iron-Product Specifications-Circuit Diagram PDF File
04-Hand steam iron-Assembly drawing PDF file
05-Handy Irons-Parts List Microsoft Excel Macro Worksheet

Fig. 5 Example of an input document for the "smart risk pre-review system"—technical regulation

The "Smart Risk Pre-review System" uses AI Deep Learning API The import history of importers, parts list, assembly drawings and circuit diagrams of the Handis Timber, among the four statutes, including those subject to AI Deep Learning variables, concluded that the applicable laws were the "wave law" and the "safety management of electrical and household goods. "The "Smart Risk Pre-review System" utilizes the AI Deep Learning API and concludes that the standard under the "transmission method" is equivalent to "KC-compliant registration" and that the test criteria are KN14-1, KN14-2. However, reaching a conclusion on the "Electrical and Household Goods Safety Management Act" requires more cloud and database information. Deep learning is carried out with data on the manufacturer's production history, quality system information, certification of applied parts and safety as AI Deep Learning variables. The required information will be provided in the cloud or in government or certification authority big data using data mining as an AI deep learning variable (feature). The more search learning about components in the "Smart Risk Pre-review System", the more advanced AI deep-learning variables will be added, and the more accurate AI deep-learning performance will be carried out on technical regulations and standards.

3.3 Designing Machine Learning AI for Conformity Check and Quality Labelling Document

The "Smart Risk Pre-review System" analyzes reports and certificates for products. It is going to design AI machine-learning algorithms to check if products were evaluated for suitability. It is designed to analyze big data in the cloud to verify reports and certificates, or to analyze and validate job-input documents. The product's model name, certification number, manufacturer, importer, and vendor information are provided as AI machine running variables and designed for analysis. AI Machine Learning search is a process to verify that compliance work has already been completed. It is required to verify compliance by entering a test report or certificate. Designs are also intended to display relevant technical regulations after checking compliance. The technical regulations legislate items that require the product to disclose to consumers. Technical regulation items are designed to provide documentation displays automatically by the "Smart Risk Pre-review Systems" based on the content of reports and certificates. The input data required for AI to analyze the task is shown in Fig. 6.

Fig. 6 Example of the input document of the smart risk pre-review system—validity check

4 Conclusion

In this paper, we studied a "smart risk preliminary review system" that can provide necessary information for market entry in response to non-tariff barriers. This system evaluates the suitability of products and analyzes big data.

The function of this system is to determine the suitability of the product and to analyze a Big Data to provide a solution. Because it is operated by cloud computing, it performs all functions non-face-to-face.

It is designed to connect the information on the production, manufacture, distribution, and customs of trade products using block chain. The information generated by the" Smart Risk Pre-Review System" analyzes big data in a database stored in the cloud. With AI relearning and AI machine learning, trade product information is designed to be transported and operated in cloud computing networks.

The "Smart Risk Pre-Review System" applies to product certification, environmental certification, hazard certification, and quality system certification. The HS code search for trade product sales requirements uses AI machine learning, and the technical rule search and standard search are designed to use AI deep learning algorithms. In addition, the conformance check AI machine learning will be learned, and the conformance DB will be designed to apply the labeling rules.

References

1. Kim BW, Kim JH (2019) Progress and impact of US-China trade friction. KIET Ind Econ Rev Spec 1–3
2. Yuri D, Erdal O, Choi MK (2019) Cybersecurity: attack and defense strategy, 230–231
3. Nakamoto S (2008) Bitcoin: a peer-to-peer electronic cash system. Available: http://bitcoin.org/bitcoin.pdf
4. Son BH, Kim JH, Choi DH (2017) Major science and technology innovation policy tasks for responding to the fourth industrial revolution. KISTER Issue Paper. Available: https://www.kistep.re.kr/c3/sub3.jsp?brdType=R&bbIdx=11289
5. Ahn SH, Han KH, Ahn SS (2002) Internet Key exchange based on PKI in wireless environment. Korean Inst Inform Sci Eng 29(2III):139–141
6. Lim JH (2018) Characteristics of blockchain technology, industry utilization and market forecast. Daehan Assoc Bus Adm 21
7. Lee JH, Youn SW (2019) Study on smart contract in blockchain. Korean Inst Inform Sci Eng 2091–2093
8. Jee EK, Jung YL, Yoo JH, Bae DH (2019) Simulation and formal verification of blockchain consensus algorithms: a case study and analysis. Commun Korean Inst Inform Sci Eng 37(4):37–48
9. Park CH, Kim MK, Kim HW (2018) A Study on building blockchain network and decentralized application development based on Hyperledger Fabric. In: Proceedings of symposium of the Korean institute of communications and information sciences, pp 5–6
10. Ministry of science (2017) ICT and future planning and ministry of justice. Electron Doc Commentary 9
11. National IT industry promotion agency (2011) Electronic document security management guidelines research report, pp 13–14

AI Analysis of Illegal Parking Data at Seocho City

Donghyun Lim and Deawoo Park

Abstract The CCTV surveillance center (below to "Center") in Seocho City Office operates 3724 CCTVs. CCTV control is operating infrastructure such as 40 Gbps self-fiber-connected network dedicated information and communication, 7 PB SAN storage and 200 virtual machines to ensure the safety of citizens. Center grow up the newest technique from now on. However, in view of Bia Data for illegal parking and stopping judgment in Center relies on the old automatic software. In this paper, we characterize images by local resident, neighborhood, site-specific, site-specific-direction, day of week, hour of day and analysis of Bia Data for judging illegal parking. We design an AI machine learning system that links the vehicle's number recognition algorithm for illegal parking. Finally we design a advanced system that analyzes the status of illegal parking and stopping judgment in Seocho City Office, where Big Data and AI are connected using spatial information and AI.

Keywords CCTV · Deep learning · Video analysis · Illegal parking · Parking stop judgement

1 Introduction

The 4th Industrial Revolution technology is being applied to the real life of citizens. 5G smartphones receive information on the movement of patients with confirmed COVID-19 in Seocho City. 5G smartphone users can receive information from COVID-19 diagnostic tests, as well as phone connections with real-time map search and AI voice recognition and voice commands to avoid infection routes. While

D. Lim · D. Park (✉)
Department of Convergence Engineering, Hoseo Graduate School of Venture, Seoul, Korea
e-mail: prof_pdw@naver.com

D. Lim
e-mail: ji3808@naver.com

© The Author(s), under exclusive license to Springer Nature Switzerland AG 2021
J. Kim and R. Lee (eds.), *Data Science and Digital Transformation in the Fourth Industrial Revolution*, Studies in Computational Intelligence 929,
https://doi.org/10.1007/978-3-030-64769-8_13
165

driving a vehicle, 5G smartphones take into account GPS and road traffic as well as video calls, and AI searches for the optimal vehicle driving route with a map and guides the vehicle with voice.

It is the traffic problem and garbage problem that Seocho City receives a lot of civil complaints for urban management. In traffic problems, especially illegal parking, which obstructs vehicle flow, is the most common complaint. To solve these traffic problems, Seocho City installed parking control CCTVs and expanded and is operating 348 parking control CCTVs in August 2020.

However, in the process of improving citizens' life satisfaction and solving illegal parking, parking enforcement CCTV has not disclosed the enforcement process data. In addition, even after the illegal parking crackdown, only individual crackdowns are notified. Only simple information such as installation location is open on the website of Seoul and Seocho City as public parking control information [1, 2].

It is necessary to utilize the advanced technologies of the 4th Industrial Revolution to improve the efficiency of the CCTV operation for parking control in Seocho City. Parking control CCTV, introduced in Seoul city and autonomous districts since the late 2000s, has introduced some artificial intelligence machine learning supervised learning concepts, but in order to satisfy the actual parking control problem, the introduction of artificial intelligence deep learning is necessary. In other words, when determining the number of cases of parking enforcement CCTV analysis, reading, and administrative disposition, artificial intelligence designed a convolutional neural network of deep learning and an artificial intelligence algorithm for detection and optimal judgment to analyze actual parking enforcement CCTV and illegal parking. It should be applied to the crackdown.

AI-based vehicle discovery and vehicle number acquisition research are in progress like **Vehicle license plate area detection using artificial intelligence deep learning** [3], **license plate recognition using polynomial-based RBFNNs** [4], **vehicle number recognition using data expansion and CNN algorithm** [5]. The parking control method using artificial intelligence deep learning is a ReID technology for tracking criminal vehicles [6, 7] can be used. It also fits the government's policy direction, which understands the 4th Industrial Revolution technology as improving administrative services to the public [8].

In this paper, after examining the actual parking control business process of Seocho City Office, we apply a binary machine learning method based on artificial intelligence supervised learning in the analysis and operation method of illegal parking control CCTV. In big data analysis, information on the parking control process in Seocho City is analyzed from the perspective of dong, time, day and system resources. In addition, for the efficiency and improvement of illegal parking enforcement in Seocho City Office, a parking enforcement method using artificial intelligence deep learning is designed and proposed.

2 Vehicle Number Recognition System Analysis

2.1 Data Preprocessing for Seocho City Vehicle Number Recognition and Extraction

In the case of vehicle number recognition using a contrast difference, the distance between the camera and the license plate, weather and shadows are greatly affected. So, in most cases, it is used **the region binarization** [9] and **the morphology technique**(open and close operation) [9]. Through this pre-processing, image quality improvement, lighting correction, and shadow distortion can be compensated.

2.1.1 Locally Adaptive Thresholding Method

Images captured in the field with a CCTV imaging device (CMOS) are difficult to identify images due to various lighting interferences. Therefore, the binarization technique is used to cancel unwanted illumination interference by using the contrast of the monochrome image.

There are many types of binarization. There is the simplest binarization method, **global fixed thresholding method** that binarizes based on a threshold value, but it is not easy to specify a license plate or vehicle number using this method alone. This is because the binarization success rate is high where the environment is prepared, but the possibility of binarization failure is high outdoors where there is a lot of lighting interference. Another method is **Locally adaptive thresholding**, which is used for calculation by subtracting a constant from the brightness average for each pixel of the image and calculating a threshold value like Fig. 1. This method is common [4].

Fig. 1 Locally adaptive thresholding (Left) and global fixed thresholding (Right)

Fig. 2 Result of opening plus closing calculation [Before (Left), After (Right)]

2.1.2   Image Morphology Technique

Since the binarized image is noisy, it is impossible to process it without correction. The method used at this time is the morphology technique. This technique maintains the characteristics of the image and processes only the shape change. Morphology is a concept similar to a mask and refers to a method of selecting pixels in an image in various shapes. Expansion, erosion, closing, and opening are possible as a method of calculating this, but generally in Fig. 2, opening and closing operations (Closing) removes noise and spots [4].

2.1.3   License Plate and Number Extraction

The outline of the binarized image in the previous section is displayed using **findcontour** among various functions of **OpenCV**. At this time, the license plate is determined using the unique ratio index of the license plate, and the number is then extracted by reflecting the ratio. If the ratio of the extraction number is not appropriate, the previous process is repeated to continue the process of finding the right license plate and number. Figures 3 and 4 is a picture of license plate and number extraction using **findcontour** function [4]. Various technologies have been developed to remove license plates and number distortion according to the shooting angle [10].

Fig. 3 Plate detection using by **findcontour** function [Outline formation (Left) and Plate box formation (Right)]

Fig. 4 Number detection using by **findcontour** function [Outline formation (Left) and Number box formation (Right)]

3 AI Machine Learning Analysis on Illegal Parking in Seocho City

3.1 Process Analysis for Illegal Vehicle Number Recognition

AI machine learning supervised learning expresses the process of recognizing the number of illegally parked vehicles in Seocho City. In order to increase the efficiency of parking enforcement, the performance of each process for recognizing the vehicle number is important. Only by improving the individual performance of each process can be expected to improve the license plate recognition rate.

Seocho City parking enforcement work process is shown in Fig. 5. Parking control CCTV attempts to recognize illegal parking vehicle license plates based on the same principle as AI machine learning supervised learning according to the already entered schedule. The CCTV control center manager visually checks the license plate information for illegally parked vehicles that were automatically cracked down the day before, deletes personal information, corrects errors, confirms the enforcement information, and issues a notice.

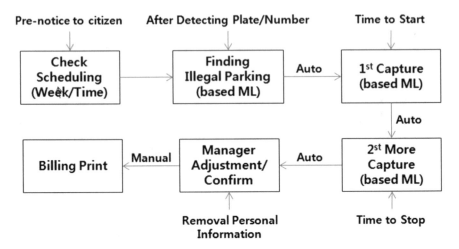

Fig. 5 Illegal parking work process

- *SW analysis design for AI machine learning*

The parking control program installed on the Seocho City parking control host described in Fig. 7 checks the pre-registered schedule, checks whether parking is available date/time, and performs a preprocessing process of binarizing by controlling the camera. After finding the feature point, extracting the license plate and vehicle number in order, record it as a target for enforcement, and take pictures in wide view and narrow view (this is called 1st capture). It moves to the next preset defined in the setting to perform the vehicle detection process, and after a certain time elapses after the first site crackdown, it returns to the memorized angle of view and re-shoots to confirm the crackdown (this is called 2nd capture).

The currently used technology uses **OpenCV**-based supervised learning engine, Fig. 6, a more advanced AI machine learning algorithm is designed to operate as an AI machine learning supervised learning process.

3.2 AI Machine Learning for Illegal Parking Enforcement

Table 1 applied to Seocho City parking enforcement 348 CCTVs carried out 869,913 cases and 246,905 cases of first and second crackdowns for one year, of which only 10.427% of the first crackdown standards were actually fined.

- *Learning AI machine learning by dong in Seocho City*

There are 18 dongs in Seocho City. Figure 7 represents the number of CCTVs installed in each building and the number of crackdowns in each building. It can be seen that the number of CCTVs is low in places with large-scale apartment

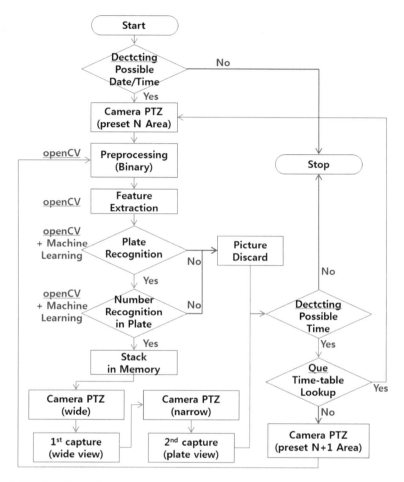

Fig. 6 Illegal parking software process

Table 1 Operating result review

Category	CCTV amount	Amount after 1st detecting	Amount after 2nd detecting	Amount of fault	Amount of judgment
Total	348	869,913	246,905	156,207	90,705
Rate (Upper/1st)	—	—	28.383%	17.957%	10.427%
Rate (Upper/2nd)	—	—	—	63.266%	36.737%

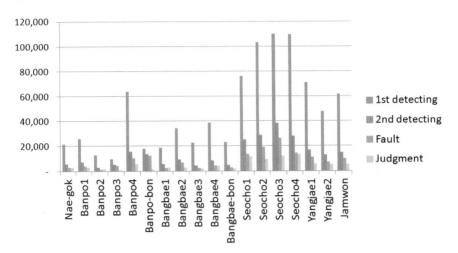

Fig. 7 Each step's amount of illegal parking detection in 18 small province

complexes, and the number of CCTVs is high in places where shopping centers and offices are concentrated. It can be seen that the number of crackdowns is high in Seocho1-dong, Seocho2-dong, Seocho3-dong, Seocho4-dong, where there are many floating populations such as Gangnam Station and Express Terminal, and Banpo4-dong and Jamwon-dong. The peculiar point is that in Banpobon-dong, the amount of enforcement is small, but the detection error rate is low.

Figure 8 expresses the value obtained by dividing the number of crackdowns per

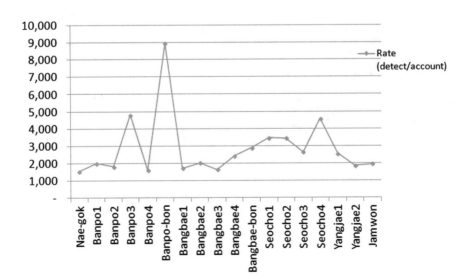

Fig. 8 Result of [Amount of 1st/Amount of CCTV] in each small province

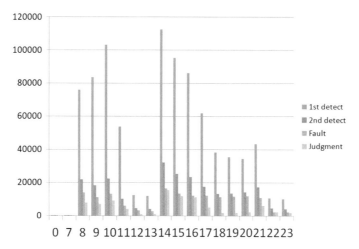

Fig. 9 Illegal parking amount comparison of daily timeline

dong by the number of cameras per dong based on the first crackdown result. You can see where the probability of finding a parked vehicle is high.

- *AI machine learning results by time*

The parking enforcement time in Seocho City is from 08:00 to 24:00. From 12 o'clock to 14 o'clock, it is a grace time according to the meal time, and illegal parking is not regulated except for a few places that are very fatal to traffic Fig. 9. Crackdowns occur during the most active hours before and after lunch.

- *AI machine learning results by day of the week*

Comparison of the number of parking enforcement by day of the week Fig. 10 shows that there is a strong tendency for illegal parking on Monday morning, Tuesday afternoon, and Thursday evening. Comparison of the number of parking enforcement on weekends Fig. 11 shows that there is a higher possibility of causing traffic jams due to illegal parking on Saturdays than on weekdays, and less than 30% of illegal parking on Sundays than on weekdays.

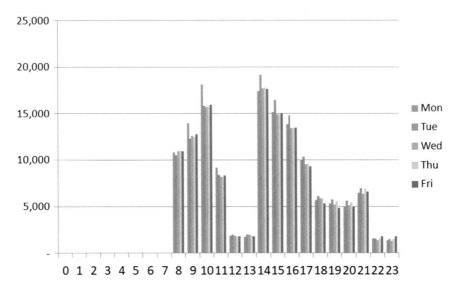

Fig. 10 Illegal parking amount comparison of week

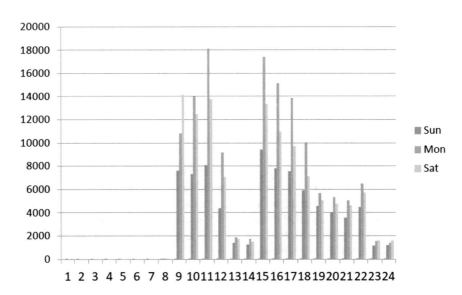

Fig. 11 Illegal parking amount comparison of sun, mon and sat

4   AI Deep Learning Design to Improve Illegal Parking Enforcement Efficiency

4.1   AI Deep Learning System Design

It analyze and design AI deep learning algorithms and methods to make illegal parking enforcement and administrative disposition more efficient. For the maximum use of AI deep learning server resources, all servers have to adjust to virtualization based on hypervisor and operates **thin provisioning**.

4.2   AI Deep Learning Process Improvement Design

The current method of analyzing illegal parking enforcement video information and operating some AI supervised learning machine learning-based engines has many inefficient factors such as waste in process management and raising the question of administrative measures for enforcement. Illegal parking enforcement schedule is managed by AI supervised learning, but illegal parking enforcement vehicle and vehicle number recognition is Fig. 12, it is suggested to design with AI deep learning.

4.3   AI Deep Learning Algorithm Application Design

Supervised learning machine learning-based algorithm, which is a method of illegal parking control system, is shown in Fig. 13, the AI deep learning algorithm DNN (Deep Neural Network) is converted to improve the performance of the entire illegal parking process.

   In AI Deep Learning DNN Algorithm, HL (Hidden Layer) 1 inputs illegal parking status, vehicle number recognition, elapsed time calculation for schedule, real-time status information, etc. as main functions to operate AI deep learning system and illegal parking control system Improves performance.

4.4   Improvement Effect When Applying AI Deep Learning

When AI deep learning and DNN algorithm are applied to illegal parking enforcement, improvement effects as shown in Table 2 can be obtained. Through AI deep learning and DNN algorithm, it can be applied to buildings where illegal parking enforcement and vehicle number recognition efficiency is low, thereby improving parking enforcement efficiency. In addition, by operating the deep learning DNN algorithm according to the number of cases in the field, the administrative power

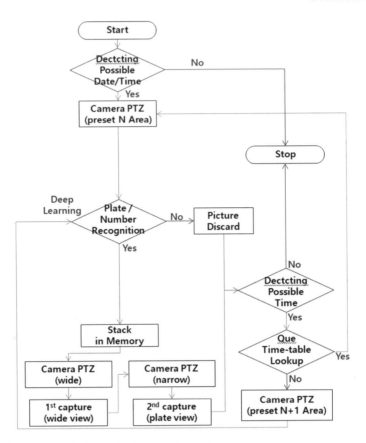

Fig. 12 Improvement design for AI deep learning SW process

Fig. 13 Improvement design
for AI deep learning process

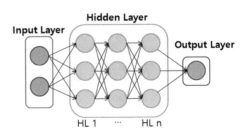

of enforcement and prevention can be concentrated in the days and times of high frequency of illegally parked vehicles. The final goal, the 4th Industrial Revolution technology, can be reflected in administrative policy to increase the safety and satisfaction of citizens.

Table 2 Comparison of AI machine learning versus AI deep learning

Category	AI machine learning (ML)	AI deep learning (DL)	Improvement effect
License plate recognition discrimination amount	100 ch × 3 frames per second	ML × 4 times (Parallel processing)	Infrastructure efficiency improvement
Number of license plate recognition learning times	Manual learning in 1–2 times/month	DL algorithm self-learning	Labor cost reduction and improved accuracy
Illegal parking control judgment method	Administrator visually checks and enters the ML	No intervention of high reliability algorithm by managers	Administrative power reduction by preemptive exception handling and labor cost reduction

5 Conclusion

When operating illegal parking control CCTV, some methods of detecting vehicles and recognizing vehicle numbers are operated based on AI machine learning supervised learning. However, inefficient SW/HW operation, location/time/day of the week, etc., the classification and regression methods of machine learning supervised learning are not effectively reflected. Therefore, there is a problem in that a lot of effort and time are consumed by the administrator due to the uniform illegal parking control method.

In this paper, we presented big data analysis by reflecting situations such as SW/HW operation and winter/time/day of the week through classification and regression methods of AI machine learning supervised learning. In addition, AI deep learning DNN algorithm was designed based on the AI learning method results by analyzing AI deep learning in order to efficiently control illegal parking vehicles. In addition, in HL (Hidden Layer) 1 using the DNN algorithm, the AI deep learning system is operated by inputting illegal parking status, vehicle number recognition, elapsed time calculation for schedule, real-time status information, etc. as main functions. An improvement plan was designed to improve performance. By reflecting the AI deep learning system design, AI deep learning process improvement design, and AI deep learning algorithm applied design, the amount of illegal license plate recognition is increased by 4 times, the number of license plate recognition learning is reduced, the reliability of illegal parking enforcement judgment is improved, and real-time data mining The improvement effect on the possibility of report was confirmed.

In the future, research on algorithms that can be applied to more than 10 illegal parking control deep learning process variables and research that can apply deep learning to disaster situations and crime prevention using illegal parking control CCTV is needed.

References

1. Seoul City parking violation control inquiry system. https://cartax.seoul.go.kr/
2. Seocho City parking enforcement system. https://traffic.seocho.seoul.kr/
3. Jeong Y, Ansari I, Shim J, Lee J (2017) A car plate area detection system using deep convolution neural network. J Korea Multimedia Soc 20(8):1166–1174
4. Kim S, Oh S, Kim J (2015) Recognition of vehicle license plate using polynomial-based RBFNNs. The Korean Institute of Electrical Engineers, pp 1361–1362
5. Lee G, Kim Y, Lee D, Kang H (2020) Vehicle number recognition using data extension and CNN algorithm. Korean Society of Surveying, Geodesy, Photogrammetry, and Cartography, pp 237–240
6. Lee S, Chu S, Kwon K, Cho N (2017) Pedestrian detection and re-identification for intelligent CCTV systems. J Korean Inst Commun Sci 34(7):40–47
7. Chae S, Choi H, Kim I (2018) Person re-identification in multiple camera environment. Inst Electron Inform Eng 380–381
8. Ministry of the Interior and Safety; Public Service, Fly with A Digital Technology, Ministry of the Interior and Safety, pp 180–186
9. Kim D, Moon H (2019) Implementation of pre-and-post processing algorithm to improve LPR (License plate recognition). Trans Korean Inst Electr Eng 68(12):1594–1600
10. Kim J (2011) Distortion invariant vehicle license plate extraction and recognition algorithm. Korea Contents Assoc 11(3):113–120

Regularized Categorical Embedding for Effective Demand Forecasting of Bike Sharing System

Sangho Ahn, Hansol Ko, and Juyoung Kang

Abstract The value of sharing economy services is increasing every year, and demand forecasting based service operations are essential for sustainable growth. For effective demand forecasting, this study proposes a categorical embedding based neural network model. The performance of this model is better than the traditional one-hot encoding based prediction; however, there are difficulties in creating a generalized prediction model due to the possibility of over-fitting of training data. Accordingly, it is possible to predict optimal demand by showing regularized performance applying techniques such as Batch Normalization, Dropout, and Cyclical Learning to the neural network. This methodology is applied to the Bike Sharing System to forecast bicycle rental demand by stations. In addition, in order to use the characteristics of global learning categories, uniform manifold approximation and projection (UMAP)-based dimensionality reduction technique is performed on the embeddings. The dimension-reduced embeddings are projected on the coordinate plane and used for K-means based cluster analysis, thereby providing an effective analysis result for demand patterns.

Keywords Sharing economy · Bike sharing system · Demand forecasting · Categorical embedding · Uniform manifold approximation and projection (UMAP)

S. Ahn · H. Ko · J. Kang (✉)
Department of e-Business, Ajou University, Suwon, Republic of Korea
e-mail: jykang@ajou.ac.kr

S. Ahn
e-mail: ash1151@ajou.ac.kr

H. Ko
e-mail: dnjsvkdls@ajou.ac.kr

© The Author(s), under exclusive license to Springer Nature Switzerland AG 2021   179
J. Kim and R. Lee (eds.), *Data Science and Digital Transformation in the Fourth Industrial Revolution*, Studies in Computational Intelligence 929,
https://doi.org/10.1007/978-3-030-64769-8_14

1   Introduction

Over the past few years, the emergence and development of a sharing economy platform has transformed the way people "share" a single public good, beyond the conventional "rental" concept of lending private property; as a result, shared items have appeared in various markets. According to PwC, a global survey agency, the sharing economy market is expected to grow from $15 billion in 2014 to about $350 billion in 2025. However, since economic services are based on sharing limited resources, optimization of service-demand by intelligent systems is a strategic task [1]. Specifically, the example of Bike Sharing System involves active rentals with several rental stations. In order to ensure that the service operates smoothly, it is necessary to analyze the bicycle rental pattern and model it so that rental stations use it to relocate bicycles. In fact, Fig. 1, based on about 9000 opinions on the citizen opinion bulletin board, confirms that there were complaints of relocation in Seoul's bike sharing system.

On the other hand, demand forecasting can lead to robust results by applying a traditional regression or tree-based model to tabular data comprising a combination of continuous and categorical variables. These methodologies have the advantage of determining feature importance, so that it is possible to find out which variables have a significant influence on demand and which variables are unnecessary [2, 3]. However, since the categorical variable was converted into a one-hot encoder to

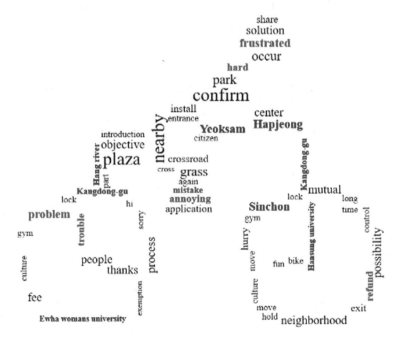

Fig. 1 Wordcloud of the bulletin board from Seoul citizens' opinion

generate a model, the relationship between the categories could not be confirmed, and it was not suitable for explaining similar rental patterns. In addition, in the bike sharing system, which has a different distribution of rental demand for each rental station, it is necessary to use multiple forecasting models for each rental station rather than a universal model; thus, the entire rental pattern cannot be implicated in each model.

Therefore, in order to solve these problems while meeting the task of forecasting demand by rental stations, this study proposes a neural network-based demand forecasting methodology applying categorical embedding for the bike sharing system. Specifically, it uses not only continuous variables such as climate information but also categorical embedding variables such as time information to train a neural network for each rental station. The study method shows that this is an excellent technique for forecasting rental demand. In addition, cluster analysis is performed to classify similar rental stations by embedding rental station information using the same methodology extensively. Unlike existing traditional practices, it will enrich demand forecast that is insufficient for each rental model by using rental embeddings and clusters that implicate the global trend of total rental data. Moreover, since multiple regularization techniques are used in the model generation and training stage, generalized results can be expected without overfitting.

The overall study contribution is as follows.

(1) *We propose a regularized categorical embedding (RCE) methodology for effective embedding of categorical variables and generalized prediction performance.*

(2) *It shows that the model's prediction performance is superior to others, which are recognized to have excellent performance in the existing tabular data.*

(3) *Categorical embedding, dimensionality reduction, and clustering methodologies are combined to find similar demand patterns.*

(4) *Effective relocation is possible in shared economic services, so it is possible to provide effective services to users.*

2 Related Works

2.1 Demand Forecasting for Bike Sharing Systems

Over the past few years, research on predicting demand for shared bicycle rental services has been ongoing. First, attempts were made to add several variables for prediction. Reference [4] generated regression equations using "gu" variables (which are administrative areas) to predict bicycle demand for specific regions. According to [5], most existing studies on demand forecasting have overlooked the variables of weather change but suggested that these variables significantly influence demand forecasting. Reference [6] conducted time series analysis using the Holt-Winters method and performed demand forecasting.

There are trends to apply various machine learning techniques. For instance, [7] used various techniques such as Ridge linear regression, support vector regression (SVR), random forest, and gradient boosted trees based on shared bicycle rental data in 2011 and 2012 in Washington, D.C. Similarly, [8] used Recurrent Neural Networks (RNN) to solve imbalanced placement of shared bicycles by rental locations and conducted research on evenly relocating rental bicycles based on shared bicycle data in New York City. Among them, a remarkable study is by [9], which used a station-centric model to predict the rental demand of bike sharing systems. In addition, a methodology was used to increase the performance by extracting global features and clustering adjacent rental stations by creating a City-Centric model that aggregates the entire station data. In this study, we adopted the city-centric technique to implicate the global pattern.

However, there is a limitation that such a study on bicycle sharing does not completely solve the problem of demand imbalances by rental stores. In most studies, it was difficult to expect accurate demand predictions to meet the latest trends using only user rental data for about a year or so. In addition, existing studies that performed daily and monthly analysis, not time-based analysis, could not determine when time zone relocations should occur. Finally, there was a limitation that the one-hot encoding method for categorical variables could not be used to reveal the deep relationships between categorical variables. Therefore, this study not only uses three-year data but also categorical variables for detailed time zones; therefore, it is possible to predict relocation by time zones. In addition, since it adopts a methodology for embedding categorical variables, it will be possible to understand intuitively for each variable and demonstrate improved performance.

2.2 Categorical Embedding

This study forecasts demand based on tabular data called public bicycle rental data, and an embedding method is applied to categorical data. Research on embedding has been actively conducted across various fields, especially in natural language processing [10] and recommendation system [11]. In fact, in the field of text mining, word embedding research is at the forefront. In word embedding research using artificial neural networks, [12] introduces Word2Vec, which learns the probability of a word appearing through the relationship between the surrounding context and a specific word. Using this, [13] propose a Structural Deep Network Embedding method to effectively detect nonlinear network structures and preserve global and local structures. Also, [14] shows that Deep Embedded Clustering is a method for simultaneously learning feature expressions and cluster allocations using deep neural networks and shows that Unsupervised Deep Embedding is also used in the analysis. As such, research is underway to embed information in various fields.

However, studies to solve the problem by applying deep learning to data including categorical variables have not been actively conducted [15]. These attempts were presented in two studies that performed well on Kaggle, which solved data analysis problems, and were highly regarded academically. In fact, [16] won the first place by applying a deep learning model to categorical metadata in the ECML/PKDD 15: Taxi Trajectory Prediction (I)" competition, suggesting that embedding actually affected performance. In the "Rossmann Store Sales" competition, [17] applied the methodology of embedding categorical data to demand forecasting problems, and not only secured third place but the categorical data represented in the vector dimension also improved the prediction performance. This has high value as an example of interpretability and transfer learning in that the meaning of the relationship between the data in each category is expressed together. The above embedding case studies show that academic and practical results in deep learning models can be applied to table data by representing categorical data in vector dimensions; however, there were limitations in overfitting because they did not deal with normalization. In addition, a specific utilization plan for the relationship between learned embeddings of variables have not been suggested.

3   Proposed Method

For the purpose of this study, two processes are required for effective demand forecasting and performance evaluation of the bike sharing system, as shown in Fig. 2. The first step is to design and train a categorical embedding-based neural network, or Regularized Categorical Embedding (RCE) model, which applies regularization techniques to predict demand by rental station.

The second step is to solve the problem that a suitable model cannot be found in a specific rental station due to data imbalances by rental stations (Station), and rental embedding is solved through cluster analysis applying Uniform Manifold Approximation and Projection (UMAP) and K-means.

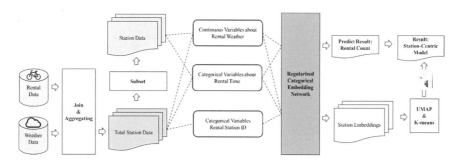

Fig. 2 Overview of our proposed model

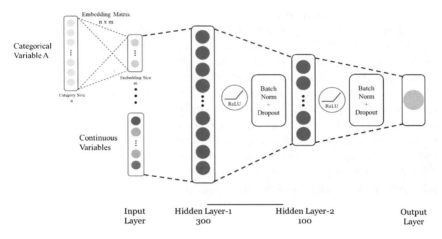

Fig. 3 Our regularized categorical embedding network

3.1 Categorical Embedding Based Network

Demand forecasting model is a regression model that predicts continuous dependent variables and must receive numerical data. Therefore, continuous variables can be used without modification; however, categorical variables must be expressed in numerical form to proceed with the analysis. There are two main methods to solve this: (a) one-hot encoding based method, and (b) embedding-based method. From an analytical perspective, embedding-based methodology is more effective as it has better predictive performance and learns the data properties [18].

The RCE method was also used for embedding categorical variables, and various normalization techniques were applied to create and distribute a practically useful rental prediction model. First, batch normalization is used to normalize the distribution of different input values for each layer. It is a representative method to improve speed, performance, and stability of the artificial neural network, and prevent overfitting during learning. [19]. Second, dropout is implemented by randomly excluding the nodes of the layer at a specific rate for each training, and not only produces the same effect as ensemble multiple models, but also prevents overfitting as in the previous batch normalization [20]. By including these techniques, it is possible to obtain a generalized prediction model without overfitting to the learning data, and to obtain well-trained embedding. The overall model architecture is presented in Fig. 3.

3.2 Model Optimization

An optimization stage was implemented in the learning stage through a pre-processed dataset, and a learning methodology termed "cyclical learning" was adopted. It is possible to reach the global minimum effectively even in small iterations by applying a policy that covers both the basic learning rate and maximum learning rate, rather than using a fixed learning rate [21]. When this methodology is used, first, a target learning rate is required. When learning for the first time, it starts from a value that divides the learning rate by a designated factor and gradually increases to the target learning rate. After achieving the target learning rate, it gradually decreases to the learning rate that progressed first time and concludes learning to complete one cycle. The loss function used for optimization is a mean squared error (MSE) and the Adam Optimizer is used by the optimizer.

3.3 Demand Forecast and Station Embedding

The above RCE methodology is applied to the data for each rental location to generate a demand forecast model. First, the reason for creating a model for each rental place, rather than a universal model based on the entire data, is that previous studies confirmed that a different number of units and rental patterns exist for each bicycle rental location. In order to confirm this, model generation for each rental shop is completed and the evaluation indicators are displayed as a histogram to examine the distribution.

The advantage of the RCE methodology is that even if a new categorical variable is included, it does not add as many dimensions as one-hot encoding; besides, it is easy to train the new categorical variable to reflect this tendency. Therefore, data of the previous rental-specific models were merged, and the rental station itself was added as a variable for embedding information in the rental station. Therefore, it is possible to learn the tendency for overall rental demand, that is, the pattern, and the entire rental shop learns this tendency and is represented by multi-dimensional continuous variables.

3.4 Projection Embedding via UMAP and Clustering Analysis

All learning categorical variables are represented by different multi-dimensional embedding vectors, which preserves a wealth of information, but has the disadvantage of being complex and difficult to interpret. Therefore, several techniques to visualize these multi-dimensional representations using two-dimensional reduction techniques were applied [22]. Initially, methods such as Principle Component Analysis (PCA) or t-SNE were used in dimensionality reduction methods, but a UMAP method that can compress multi-dimensional information more efficiently and describe the global structure has been proposed. Indeed, [23] projected movie information or user information acquired in the recommendation system in two dimensions using

UMAP and showed that this is more effective than other techniques. Therefore, UMAP dimensionality reductions can be applied to the embedding vectors learned through the RCE technique, and each one can be visualized and expressed.

In this study, UMAP was used to analyze the pattern by dimensionally reducing the embedding vector that was learned earlier and clustering it. Based on this, it is possible to effectively represent information on rental embedding vectors that learned global trends based on the entire data in two dimensions to create clusters between rental stations with similar rental patterns. Therefore, rental demand forecast, which was inadequate by the rental model due to unbalanced data or patterns, is supplemented with the model with highest predictive performance in the cluster, and the rental pattern is analyzed.

4 Experiments

4.1 Datasets

In this study, about 12.5 million rental datasets from January 2017 to May 2019 were collected from the "Seoul City Public Bike Usage Status" to predict the demand on a specific "rental station" and analyze time pattern for bike relocation. At this point, since there were a few different data types in the yearly dataset, they were all unified in the same format, and preprocessing was performed for missing values and outliers. In addition, precipitation, fine dust, and temperature information for each time zone were collected using the Seoul Metropolitan Meteorological Administration API to add climate information. Moreover, climate information was not subdivided into administrative regions because it was confirmed in the EDA (Exploratory Data Analysis) process that there were no significant climatic differences in each region. By combining the two types of data collected using this method, it was possible to construct the final type of data necessary for demand forecasting. In this dataset, categorical variables are time information such as rental time, rental day, rental month, season, and weekend. Continuous variables are climate information such as humidity, wind speed, temperature, precipitation, fine dust, and ultrafine dust.

4.2 Experimental Setup

This experiment was implemented through the Pytorch-based fast.ai library in Google Colab's GPU Tesla K80 environment. In order to minimize the mean squared error (MSE) loss function, the Adam-Optimizer [24], which can reach global solutions faster than other optimizers, was used. Various hyperparameters were used for learning via the RCE methodology. Most of them used the default value of fast.ai, but the learning speed was 0.04, and Cyclical Learning rates were applied to learn for

Table 1 The result of dimension with embedding and one-hot encoding

Variable Name	# of Category	# of dimension (Embedding)	# of dimension (One-hot)
Station ID	1505	96	1505
Rental hour	24	9	24
Rental day	7	5	7
Rental month	12	6	12
Rental season	4	3	4
Rental weekend	2	2	2

10 and 30 epochs. In the case of category variables, the size of the embedding must be defined. In the case of fast.ai, the size of embedding N_e is basically set by the following equation based on the number of categories N_o per original categorical variable [12].

$$N_e = round(1.6 \times N_o^{0.56})$$

Then, the result of dimensions, embeddings, and one-hot encoding method, to which the above formula is applied, are given in Table 1, respectively. It can be observed that as the number of categories increase, the embedding method represents data in a much smaller number of dimensions than one-hot encoding.

4.3 Evaluation Metrics

We tried to compare the results through R Square, which is one of the elements of performance evaluation borrowed from the traditional statistical model. In addition, to generalized performance evaluation, k-fold cross-validation was implemented. This is to derive the average of the predicted values by applying the algorithm and divide the data set by a certain number that can prevent overfitting [25]. In this study, fivefold cross-validation was performed.

4.4 Baseline

First, to evaluate the performance of the embedding model, a predictive model was created for four rental locations with a large number of rentals. Furthermore, for comparison, one-hot encoding multiple linear regression, XGBoost, and random forest models were compared, and the results are as follows. The performance of the proposed RCE model is indicated by the results of 30 and 10 learning cycles. The mentioned models are LR (Linear Regression), XGB (XGBoost), and RF (Random Forest), respectively, and the results of 100 learning cycles are indicated.

(1) *Linear Regression (LR): Linear regression is a widely used technique for predicting dependent variables based on several independent variables.*

(2) *XGBoosting (XGB): XGB is an effective and widely used machine learning method that extends beyond billions of examples using significantly fewer resources than existing systems [26].*

(3) *Random Forest (RF): Classification and regression by randomForest, the best split among all variables, is divided using the highest predicted value among each predictor variable that is randomly selected by the node [27].*

5 Results

5.1 Model Performance

Table 2 shows the performance of RCE models and those introduced in the baseline for the four rental shops with high rental volume. The results show that the embedding model has the best performance compared to one-hot encoding, as demonstrated in several studies. In addition, compared to traditional models, the performance of neural network-based models is found to be superior. The performance of the proposed NN-based models was indicated by 15 learning cycles and 100 learning cycles for LR, XGB, and RF.

Table 2 Evaluation results

	NN+Embeding (RCE)	NN+One-hot	LR+One-hot	XGB+one-hot	RF+One-hot
Hongik university station exit 2(113)	0.5558	0.5463	0.4625	0.2975	0.2929
Yeouinaru station exit 1 (207)	0.6092	0.6068	0.3710	0.2904	0.2857
Ttukseom amusement park station exit 1 (502)	0.6083	0.5993	0.4318	0.3215	0.3197
Mapo-gu community sports center (152)	0.5326	0.5270	0.3835	0.3562	0.3545

As the results show, the performance of forecasting the RCE methodology with embedding in categorical variables was the highest under the same conditions. However, even the best-performing RCE model did not perform well in all rental-specific models. Through this, we can see that R Square had a similar pattern as regular distribution, and rental shops had difficulty in finding the pattern with only rental models.

5.2 Station Embedding and Clustering

We compared the performance of the RCE methodology and the baseline at the top four rental stations with the highest volume of rental history and found that the RCE methodology was the best performer, but there were rental stations that were also difficult to apply. Therefore, to solve this problem, we tried to embed the rental station itself as a variable, and the result is illustrated in the following figure. This enables the following simple distance formula to calculate the similarity between 1500 rental stations.

$$distance = \sqrt{(e_{a_1} - e_{b_1})^2 + \ldots + (e_{a_n} - e_{b_n})^n}$$

A bigger advantage is that cluster analysis can be performed by dimensioning multidimensional embedding vectors. As mentioned earlier, this study has reduced the 96 dimensions of the rental station to two dimensions by applying UMAP. The reduced dimensions are known to be in the best form to describe the entire data structure. Therefore, a cluster analysis using K-mean was performed based on the reduced two dimensions. Figure 4 produced clusters with four rental patterns through

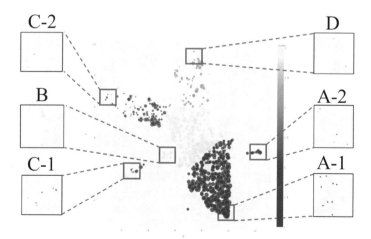

Fig. 4 Station embedding scatter plot colored by cluster using K-means

K-means, and models with R Square values below 0.3 can be used practically, such as replacing them with models with values of R Square in clusters greater than 0.3.

5.3 Rental Station Pattern Analysis: Time Information

The first pattern in the rental cluster relates to time information. As shown in Fig. 4, a group of six rental locations with well-characterized features in four clusters was sampled to analyze the number and pattern of rental time variables, and the analysis results are illustrated in Fig. 5.

The nine rental stations located in A-1 group observed distinctive patterns of time zones. The existence of such a clear pattern confirms that a model with relatively good predictive performance is generated. On the other hand, even if they belonged to the same group A, the five rental stations located in A-2 group could not confirm the clear pattern, but the remaining rental stations, except "Cheonggyecheon Yeongdogyo Bridge," had a small number of platforms. In common, it was found that it was a rental station with very low number of rentals. Four rental stations belonging to cluster B always had a flat rental, but it was confirmed that there were patterns of more than 700 rentals at 8 am. This related to work time, and it can be assumed that many office goers use public bicycles at this time of the day. Next, six rental stations, belonging to group C-1 of cluster C, had rental patterns by time zone between the rental centers except "Eunpyeong Arts Center" rental station. It was confirmed that the average number of rental hours during the day was much higher than at other rental locations. It was confirmed that five rental stations belonging to C-2 group had a pattern similar to A-1 group, but an incomplete pattern. Finally, it can be confirmed

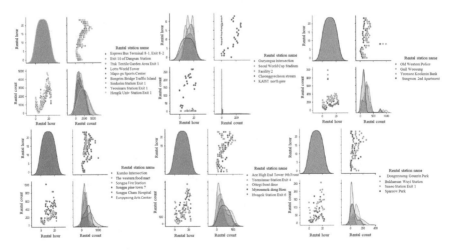

Fig. 5 Analysis of rental patterns over time, groups A-1, A-2, B from top left, and groups C-1, C-2, D from bottom left

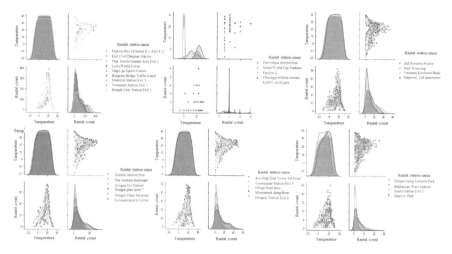

Fig. 6 Analysis of rental patterns over temperature, groups A-1, A-2, B from top left, and groups C-1, C-2, D from bottom left

that incomplete rental patterns, such as C-1 and C-2, exist in the four rental stations belonging to cluster D.

5.4 Rental Station Pattern Analysis: Temperature

The second pattern in the rental community relates to temperature. Similar to the previous analysis, six groups of rental stations were sampled from four clusters to analyze the number of rental patterns and the pattern of temperature variables, and the analysis results are shown in Fig. 6.

First, the nine rental shops belonging to A-1 group clearly showed a pattern with the highest number of rentals when the temperature was about 25°. On the other hand, the rental pattern was not clearly revealed in the four rental offices belonging to A-2 group, which was confirmed because of the small number of rentals. The four rental stations belonging to cluster B, similar to A-1 group, showed a clear rental pattern according to the temperature, confirming that active rentals occurred when the temperature was around 25°. Next, C-1 group and C-2 group revealed clear patterns, similar to A-1 group and B group, which demonstrated good performance unlike the previous time zone. However, it is confirmed that there were incomplete patterns in the four rental stations belonging to D cluster.

6 Conclusion

In this study, we propose a regularized categorical embedding methodology that can learn not only the station-centric model but also the overall trend to predict the rental demand for shared bicycles. Our model shows better predictive performance from a minimum of 16% to a maximum of 53% through embedding with non-sparse properties, while having fewer dimensions for each category in the categorical variables compared to one-hot encoding. Also, the rental station itself can be learned by embedding and combining with a dimensional reduction method based on UMAP, which can be effectively used in terms of analysis. This was an efficient way to find similar rental locations through cluster analysis and was used to describe some rental locations that were difficult to explain in the rental-oriented model. This can be confirmed through the distribution of variables of rental stations belonging to the same rental pattern cluster and EDA. The proposed methodology is expected to be an effective prediction for station-centric models with unbalanced performance distribution and the prediction of rental demand for shared bicycles.

However, this study has two limitations. First, the high performance RCE model for each rental station needs to train a new model for each station, so the deeper the layer and the larger the number of rental stations, the higher the cost of the neural network learning. Second, the effectiveness of the cluster analysis was evaluated qualitatively, without numerical proof, thereby limiting its precise use. Therefore, such problems must be developed to reduce the computational cost by transferring the learned embedding variable, and to prove the similarity by fitting the cluster-specific pattern as a linear function.

Acknowledgements This research was supported by the MSIT (Ministry of Science and ICT), Korea, under the ITRC (Information Technology Research Center) support program (IITP-2020-2018-0-01424) supervised by the IITP (Institute for Information and communications Technology Promotion).

References

1. Guo L et al (2018) Quick answer for big data in sharing economy: innovative computer architecture design facilitating optimal service-demand matching. IEEE Trans Automation Sci Eng 15(4):1494–1506
2. Lindsey C, Sheather S (2010) Variable selection in linear regression. Stata J 10(4):650–669
3. Zheng H, Yuan J, Chen L (2017) Short-term load forecasting using EMD-LSTM neural networks with a Xgboost algorithm for feature importance evaluation. Energies 10(8):1168
4. Kim KJK, Choi K, Keechoo (2011) Development of regression-based bike direct demand models. Korean Soc Civil Eng D 31(4D):489–496
5. Do M, Noh Y-S (2014) Analysis of the affecting factors on the bike-sharing demand focused on Daejeon City. Korean Soc Civil Eng 34(5):1517–1524
6. Lim H, Chung K (2019) Development of demand forecasting model for seoul shared bicycle. J Korea Contents Assoc 19(1):132–140

7. Yin Y-C, Lee C-S, Wong Y-P (2012) Demand prediction of bicycle sharing systems. URL http://cs229.stanford.edu/proj2014/YuchunYin,ChiShuenLee,Yu-PoWong,DemandPredictionofBicycleSharingSystems.pdf
8. Chen P-C et al (2020) Predicting station level demand in a bike-sharing system using recurrent neural networks. IET Intell Transp Syst
9. Zeng M et al (2016) Improving demand prediction in bike sharing system by learning global features. Mach Learn Large Scale Transp Syst (LSTS) @KDD-16
10. Howard J, Ruder S (2018) Universal language model fine-tuning for text classification. arXiv preprint arXiv:1801.06146
11. Liu DC et al 2017 Related pins at pinterest: the evolution of a real-world recommender system. In: Proceedings of the 26th international conference on world wide web companion
12. Mikolov T et al (2013) Distributed representations of words and phrases and their compositionality. In: Advances in neural information processing systems
13. Wang D, Cui P, Zhu W (2016) Structural deep network embedding. In: Proceedings of the 22nd ACM SIGKDD international conference on knowledge discovery and data mining
14. Xie J, Girshick R, Farhadi A (2016) Unsupervised deep embedding for clustering analysis. In: International conference on machine learning
15. Howard J, Gugger S (2020) Fastai: a layered API for deep learning. Information 11(2):108
16. De Brébisson A et al (2015) Artificial neural networks applied to taxi destination prediction. arXiv preprint arXiv:1508.00021
17. Guo C, Berkhahn F (2016) Entity embeddings of categorical variables. arXiv preprint arXiv:1604.06737
18. Zhang K et al (2015) From categorical to numerical: multiple transitive distance learning and embedding. In: Proceedings of the 2015 SIAM international conference on data mining. SIAM
19. Ioffe S, Szegedy C (2015) Batch normalization: accelerating deep network training by reducing internal covariate shift. arXiv preprint arXiv:1502.03167
20. Srivastava N et al (2014) Dropout: a simple way to prevent neural networks from overfitting. J Mach Learn Res 15(1):1929–1958
21. Smith LN (2017) Cyclical learning rates for training neural networks. In: 2017 IEEE winter conference on applications of computer vision (WACV). IEEE
22. McInnes L, Healy J, Melville J (2018) Umap: uniform manifold approximation and projection for dimension reduction. arXiv preprint arXiv:1802.03426
23. Kang K et al (2019) Recommender system using sequential and global preference via attention mechanism and topic modeling. In: Proceedings of the 28th ACM international conference on information and knowledge management
24. Kingma DP, Ba J (2014) Adam: a method for stochastic optimization. arXiv preprint arXiv:1412.6980
25. Fushiki T (2011) Estimation of prediction error by using K-fold cross-validation. Stat Comput 21(2):137–146
26. Chen T, Guestrin C (2016) Xgboost: a scalable tree boosting system. In: Proceedings of the 22nd ACM SIGKDD international conference on knowledge discovery and data mining
27. Kam HT (1995) Random decision forest. In: Proceedings of the 3rd international conference on document analysis and recognition, Montreal, Canada

Development of a Model for Predicting the Demand for Bilingual Teachers in Elementary Schools to Support Multicultural Families—Based on NEIS Data

Jinmyung Choi and Dooyeon Kim

Abstract Education-related administrative agencies intend to analyze NEIS data that has been in operation for over 15 years by applying big data and artificial intelligence technologies to predict social issues and establish policies. Therefore, this study developed a test model using big data to analyze the data of school sites and data held by administrative agencies and applied the actual data to this model to analyze the policy's appropriateness and effectiveness. From 2013 to 2018, a learning model was created based on the number of bilingual teachers for elementary school students in multicultural families in Gyeongsangnam-do. As a result of predicting the number of bilingual teachers in 2015, there were 40, but the test model was analyzed to require 42. It was confirmed that the test model's accuracy could be further improved if more data on the learning model is accumulated. This study's results can help establish and implement policies in various educational fields through big data analysis.

Keywords Bilingual · Education · Multicultural · Analytical model · Big data · Machine learning

1 Introduction

As defined by dictionary, education is a process of teaching, training, and learning, especially in schools, colleges, or universities, to improve knowledge and develop skills [1]. According to the dictionary's definition of education, people can learn a lot through education in school and gain abilities. However, it was found that children of multicultural families in Korea have difficulty learning or show relatively lower academic achievement than children of typical families [2]. There may be several reasons for this, but the difference in speaking ability and understanding of a language

J. Choi · D. Kim (✉)
Department of Convergence IT, Konyang University, Chungchungnam-Do, Republic of Korea
e-mail: kimdoo@konyang.ac.kr

J. Choi
e-mail: jameschoi@konyang.ac.kr

© The Author(s), under exclusive license to Springer Nature Switzerland AG 2021
J. Kim and R. Lee (eds.), *Data Science and Digital Transformation in the Fourth Industrial Revolution*, Studies in Computational Intelligence 929,
https://doi.org/10.1007/978-3-030-64769-8_15

can have a significant influence [2]. Also, the role of a teacher with the ability to understand the cultures of different communities to remove cultural boundaries, to understand different issues to communicate correctly with children of multicultural families, and to increase academic success is vital [3, 4]. Thus, it is necessary to have a bilingual teacher with qualifications verified through sufficient training. However, it is difficult to determine whether the number of bilingual teachers is sufficiently secured at school sites.

Recently, intelligent services are emerging as new value creation systems, such as expanding user convenience with information services utilizing big data analysis and artificial intelligence and predicting future situations. Using these technologies, we develop a model that predicts the demand for bilingual teachers necessary for the smooth education of children from multicultural families. To develop a bilingual teacher demand model, data of bilingual teachers from the Gyeongsangnam-do Office of Education were used from NEIS, which is being used in educational administration.

Chapter 2 of this paper analyzes the learning status of children of multicultural families, explains the characteristics of NEIS, and explains the method and procedure for analyzing big data. Chapter 3 explains the process of developing a big data analysis model, and Chap. 4 explains the suitability of the analysis model developed in this study by comparing the results of analyzing the demand for bilingual teachers using an analysis model and the number of bilingual teachers actually employed in schools.

2 Related Research and Status

2.1 Academic Achievement Status of Children of Multicultural Families

Table 1 shows the current status of children from multicultural families in Korea from 2014 to 2018, and as of 2018, the number of children from multicultural families in Gyeongsangnam-do is 17,723 [5].

Table 1 Current status of children of multicultural families in the Republic of Korea by age

Year	Status by age (Persons)				
	Total	Under 6 years old	7–12 years old	13–15 years old	16–18 years old
2018	237,506	114,125	92,368	19,164	11,849
2017	222,455	115,085	81,826	15,753	9,791
2016	201,333	113,506	56,768	17,453	13,606
2015	197,550	116,068	61,625	12,567	7,290
2014	204,204	121,310	49,929	19,499	13,466

In the study of the learning status of children of multicultural families [2], three sub-areas including learning motivation, learning strategy, and learning support in the learning characteristics area, and Korean language, belonging, support in the home, understanding of father or mother culture, and social five specific areas, including relations, were investigated.

It is believed that underachieving students of multicultural families do not receive support from teachers or friends compared to other student groups. Even if the students of the same multicultural family are in sync with learning in terms of learning characteristics, there is a learning strategy, and learning support is provided, there is a possibility that they will not fall into sparse learning.

In the area related to the sense of belonging, the group with children of poor learning of the general family had a significantly higher average score than the upper-middle group of children of multicultural families. Also, it was found that the average score was statistically significantly higher in the upper-middle group of children of multicultural families compared to the underachieving group of children of multicultural families. Through this, it was found that children of multicultural families have a relatively lower sense of belonging than children of general families, regardless of whether they are sparse in learning or not. In the end, it can be seen that a sense of belonging acts as an essential factor in improving academic achievement and forming a sense of identity for children of multicultural families. Therefore, it was analyzed that it is necessary to provide various institutional supports and opportunities to increase the sense of belonging to children with sparse learning in multicultural families.

Table 2 shows the sparse learning factors for children of multicultural families from the perspective of teachers who have experience teaching children of multicultural families. Lack of communication with parents or lack of interest can be seen as the most significant factor, followed by lack of Korean proficiency, low motivation for achievement, lack of preemptive learning, and emotional and psychological

Table 2 Factors of sparse learning among children of multicultural families

Division	Ratio (%)
Lack of Korean language skills	20.5
Lack of background knowledge about Korean culture	9.4
Prerequisite learning deficit	10.8
Low intellectual ability	3.1
The low motivation for achievement	11.7
Emotional and psychological difficulties	10.3
Difficulty in forming social relationships such as peer relationships	8.0
Lack of communication with parents, lack of interest and care	22.3
Etc	4.0

instability. Since these factors of sparse learning are related to each other, if one problem is solved, several other problems can be solved. Also, since many factors occur in schools, it can be judged that the existence and role of bilingual teachers in schools are vital.

2.2 NEIS (National Education Information System)

NEIS is a nationwide education administration information system used by students, parents, teachers, and education-related public officials. It connects more than 10,000 elementary, junior high, high school, and special schools nationwide and 195 education-related administrative agencies and the Ministry of Education through the Internet [6].

The NEIS has been used since 2003, and data such as grades, curriculum activity, and non-curriculum activities in school, and activities outside of school-related to all students enrolled in elementary and secondary schools are accumulated. The system supports 92 school affairs and academic affairs, provides 56 school administration services, 102 general administration services, 18 online services for the general public, and online materials used in approximately 4.2 million college admission courses. In addition to providing public services by linking NEIS with the information systems of related organizations such as the Ministry of the Interior and Safety and the Supreme Court, it also establishes and implements education-related policies using NEIS data.

2.3 Big Data Analysis Method and Procedure

To analyze big data, a predictive analysis method for unstructured data such as unstructured text, video, and audio, which accounts for 95% of the data to be analyzed, and structured data is required [7]. The process of extracting insights from big data consists of two processes: Data Management and Analytics. The data management process includes Acquisition and Recording, Extraction, Cleaning and Annotation, Integration, Aggregation and Representation, and Analytics include Modeling and Analysis, Interpretation [8]. Another big data analysis method is descriptive analytics, inquisitive analytics, predictive analytics, prescriptive analytics, and pre-emptive analytics [2]. Big data analysis methods include descriptive analytics, inquisitive analytics, predictive analytics, prescriptive analytics, and pre-emptive analytics, as shown in Fig. 1 [9].

Inquisitive analytics is a survey to verify/reject data such as statistical analysis and factor analysis. Predictive analytics relates to predictive and statistical modeling to determine future possibilities. Prescriptive analytics is about optimization and

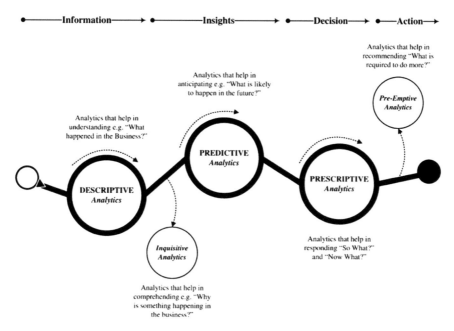

Fig. 1 Classification of types of big data analytical methods

randomization testing to evaluate how to improve service levels while reducing costs. Pre-emptive analytics can take preventive action against events that could have an undesirable impact on performance.

To analyze big data, the procedures of data collection, data cleaning, data analysis, and visualization are generally followed. The data collection step may use a method of using an open application programming interface (Open API), a web crawling, or a web scraping method. The data cleaning step is a process of purifying the collected data into a form that is easy to analyze. It is more essential for unstructured data than structured data, and text mining is mainly used. In the data analysis and visualization stage, data is analyzed through Opinion Mining, Social Network Analysis, Cluster Analysis, and Classification, and Word Cloud and pyLDAvis are used to visualize the analysis results.

3 Analytical Model Development

3.1 Model Development Process

Linear regression, one of the statistical algorithms, is widely used to analyze big data [10–12]. Big data can be effectively analyzed by applying machine learning or deep learning, so consider this and develop a data analysis model in the order shown in Fig. 2.

Fig. 2 Classification of types of big data analytical methods

- Design Items—Design optimization items (Rule Set, algorithm, features, etc.) to match the characteristics of the target model for machine learning-based data analysis
- Correction—Correction for optimization
- Data set design—Data Set (Model) design through service analysis
- Validation and optimization—Model validation and algorithm optimization through training and test data sets
- Setting parameters—Setting optimized model parameters
- Optimal model design—Optimal model design through test data.

Data to be used in the analysis model, the number of elementary schools, the number of children from multicultural families, and the number of bilingual teachers were obtained from two data sources. One was provided directly by the Gyeongsangnam-do Office of Education, and the other was obtained from NEIS. Comparing the two acquired data, it performs a refining operation for those with inconsistent or overlapping contents and those with different data formats. In order to evaluate the optimal algorithm by considering that the response variable in the optimal model is a numeric type, scikit-learn's SVM (Support Vector Machine) kernel SVC (Support Vector Classifier) provides the optimal choose an algorithm. This process is implemented with the code shown in Table 3.

Table 3 Machine learning algorithm evaluation code

```
train, test = train_test_split(df, test_size=0.3)
train_x = train.drop('multicultural', 1)
train_y = train['multicultural']
test_x = test.drop('multicultural', 1)
test_y = test['multicultural']

kernel_list = ['linear', 'rbf', 'sigmoid']
C_list = [2**-5, 2**-3, 2**-1, 2**1, 2**3, 2**5, 2**7, 2**9, 2**11, 2**13, 2**15]

for kernel in kernel_list:
    for C in C_list:
        model = svm.SVC(C=C, kernel=kernel).fit(train_x, train_y)
        predited = model.predict(test_x)
        print (kernel, C, accuracy_score(test_y, predited))
```

Table 4 Elementary school, students, and bilingual teachers by year

Year	Province	Number of schools	Number of students	Number of teachers
2018	Gyeongsangnam-do	156	639	45
2017		165	638	50
2016		157	587	43
2015		147	542	40
2014		174	417	37
2013		109	243	32

3.2 Data Set Design

Children of multicultural families whose mothers or fathers are foreigners may have lower academic achievement due to their foreign mothers or foreign fathers who are not familiar with the Korean language, as compared to other students of the same age. Thus, objective and technical means are needed to establish a reasonable budgetary basis in predicting the demand for the smooth supply and demand of bilingual teachers who will teach children of multicultural families.

This study aims to predict the number of bilingual teachers using the number of schools and students based on cumulative data on bilingual teachers from 2013 to 2018, taking elementary schools in Gyeongsangnam-do as a sample. Table 4 summarizes the data received through the Gyeongsangnam-do Office of Education and the data extracted from the NEIS.

3.3 Analysis Model Implementation

Since it is a model for predicting the number of bilingual teachers, linear regression analysis is the best explanation or prediction of the numeric dependent variable, so the data analysis model uses a linear regression analysis algorithm.

Statistical results of the final analysis result are generated using 'statsmodels' so that statistical analysis and time series analysis previously performed in R can be performed in the Python programming language.

'Statsmodels' is a Python module that provides classes and functions for the estimation of many different statistical models, as well as for conducting statistical tests, and statistical data exploration. Pandas is used to effectively process data in a two-dimensional matrix structure, such as the number of bilingual teachers for the number of children in a multicultural family, or the number of teachers in a bilingual language for the number of schools in the jurisdiction. For the statistical model, Ordinary Least Squares (OLS) was used, and R-Squared was used for model verification.

A part of the source code of the data analysis model is shown in Table 5.

Table 5 Data analysis model code

```
teacher = { 'Year': [2013, 2014, 2015, 2016, 2017, 2018],
            'number_of_students': [243, 417, 542, 587, 638, 639],
            'number_of_schools': [109, 174, 147, 157, 165, 156],
            'number_of_teachers': [32, 37, 40, 43, 50, 45]
          }

df = DataFrame(teacher,columns=
['Year', 'number_of_students', 'number_of_schools', 'number_of_teachers'])

X = df[['number_of_students', 'number_of_schools']]
Y = df['number_of_teachers']

regr = linear_model.LinearRegression()
regr.fit(X, Y)

print ('Intercept: \n', regr.intercept_)
print ('Coefficients: \n', regr.coef_)

New_number_of_students = 542
New_number_of_schools = 147

print ('Predicted number of teachers: \n',
regr.predict([[New_number_of_students, New_number_of_schools]]))
print ('y = a0*x1 + a1*x2 + b: \n', [(New_number_of_students * regr.coef_[0]) +
    (New_number_of_schools * regr.coef_[1]) + regr.intercept_])

X = sm.add_constant(X)
model = sm.OLS(Y, X).fit()
predictions = model.predict(X)
print_model = model.summary()
print (print_model)
```

4 Analysis Model Application Result

After training an analysis model using related data sets such as the number of schools belonging to the Gyeongsangnam-do Office of Education from 2013 to 2018, the number of students, and the number of bilingual teachers, the result of applying the model by inputting data from the Gyeongsangnam-do Office of Education in 2015 is R for model performance measurement. The R-squared value is 0.874, and the adjusted R-squared value shows 0.79. As the R-squared value represents a value close to 1, it can be seen that the predicted value of the number of bilingual teachers who performed linear regression analysis is converging within the expected range through data analysis.

The intercept value is 21.9439, $\beta 0 = 0.0383$, $\beta 1 = -0.0022$, and the predicted value when inputting the 2015 data of 542 students and 147 schools are 42.36321631, which is the actual number of bilingual teachers in 2015, 40.

There was a slight difference between the number of bilingual teachers and the actual number of teachers by prediction, but this is because the size of the data set used for learning is not large enough. Therefore, the prediction accuracy can be improved by reinforcing the training data set.

Tables 6 is the results of applying the data analysis model. From this, it can be seen that the big data analysis model presented in this study is significant.

5   Conclusion

As the use of big data and artificial intelligence technologies is becoming more active in various fields of society, education is also attempting to apply these technologies in various forms.

Due to internationalization, the number of multicultural families in every country is on the rise. Children of these multicultural families were found to have relatively lower academic achievement through education than those of typical families due to factors such as language, culture, sense of belonging, religion, and environment. To solve this problem, the need for bilingual teachers is very high. However, due to the lack of awareness of the importance of bilingual teachers in each region, it is difficult to secure a budget for hiring them.

To solve such a real problem, a linear regression analysis algorithm was applied, a data analysis model was developed in Python programming language using pandas and scikit-learn libraries, and the model was verified with R-Squared. The data analysis model predicted the appropriate number of bilingual teachers by entering the number of multicultural family students and the number of bilingual teachers in elementary schools belonging to the Gyeongsangnam-do Office of Education in 2015. It can be determined that the data analysis model is appropriate by deriving 42.36321631 people.

In this way, when the predicted value is derived based on the data, the need for the number of bilingual teachers can be raised to the budget department based on the objective predicted value. Through this, it is expected that the appropriate number of bilingual teachers to support children of multicultural families who have difficulties in learning and adapting to society will be maintained.

NEIS accumulates a vast amount of data, but there are not many cases of applying artificial intelligence technologies such as big data and machine learning to policy-making in administrative agencies using these data and solving various problems

Table 6 Results of bilingual teacher count prediction mode

```
Intercept:
 21.94386985246164
Coefficients:
 [ 0.03828401 -0.0022489 ]
New_number_of_student:
 [542]
New_number_of_school:
 [147]
Predicted number of teacher:
 [42.36321631]
y = a0*x1 + a1*x2 + b:
 [42.36321630911591]
                       OLS Regression Results
=========================================================================
Dep. Variable:        number_of_teacher  R-squared:                0.874
Model:                             OLS  Adj. R-squared:            0.790
Method:                  Least Squares  F-statistic:               10.38
Date:                 Mon, 11 Nov 2019  Prob (F-statistic):       0.0449
Time:                         17:47:06  Log-Likelihood:          -12.808
No. Observations:                    6  AIC:                       31.62
Df Residuals:                        3  BIC:                       30.99
Df Model:                            2
Covariance Type:             nonrobust
=========================================================================
                     coef   std err       t    P>|t|    [0.025   0.975]
-------------------------------------------------------------------------
const              21.9439    8.980   2.444    0.092    -6.635   50.523
number_of_student   0.0383    0.011   3.379    0.043     0.002    0.074
number_of_school   -0.0022    0.077  -0.029    0.979    -0.249    0.244
=========================================================================
```

arising in the educational field. Therefore, this study can be seen as meaningful as presenting an example of using the vast amount of data held by the government in the policy. The model of this study will be further developed so that in the future, it can be used by teachers, parents, and students in the process of college admission.

References

1. Oxford advanced learner's dictionary. www.oxfordlearnersdictionaries.com/definition/english/education?q=education
2. Oh SC, Chang KS, Goo YS (2013) A study on educational support for low-performing students with multicultural backgrounds: Research Report RRI 2013-2. Korea Institute of Curriculum and Evaluation
3. Banks JA, McGee Banks CA (2019) Multicultural education: issues and perspectives, 10th edn. Wiley
4. Karacabey MF, Ozdere M, Bozkus K (2019) The attitudes of teachers towards multicultural education. Eur J Edu Res 8(1):383–393
5. Ministry of gender equality and family: statistics on multicultural families
6. Ministry of education: introduction to NEIS. www.neis.go.kr/pas_mms_nv88_001.do
7. Gandomi A, Haider M (2015) Beyond the hype: Big data concepts, methods, and analytics. Int J Inform Manag 35(2):137–144
8. Labrinidis A, Jagadish HV (2012) Challenges and opportunities with big data. Proc VLDB Endowment 5(12):2032–2033
9. Sivarajah U, Kamal MM, Irani Z, Weerakkody V (2017) Critical analysis of Big Data challenges and analytical methods. J Bus Res 70:263–286
10. Genuer R, Poggi JM, Malot CT, Vialaneix NV (2017) Random forests for big data. Big Data Res 9:28–46
11. Saber AY, Alam AKMR (2017) Short term load forecasting using multiple linear regression for big data. In: 2017 IEEE symposium series on computational intelligence (SSCI), pp 1–6
12. Wang H, Yang M, Stufken J (2019) Information-based optimal subdata selection for big data linear regression. J Am Stat Assoc 114(525):393–405

Research on Implementation of User Authentication Based on Gesture Recognition of Human

Jungseon Oh, Joongyoung Choi, Kwansik Moon, and Kyoungho Lee

Abstract Biometrics technology has taken the authentication technology one step further, and the solutions, based mainly on static characteristics (fingerprint, face, retina, etc.), have been commercialized. But there is a possibility that this technology also can be maliciously bypassed. In that respect, research on new authentication technology based on dynamic characteristics (gesture, walking, keystrokes, etc.) are being underway. In the national security facilities such as nuclear power plants and power generation that require high security level, they are actively adopting biometric authentication technology based on FIDO certification standards As security breach incidents become more sophisticated and the scale of damage increases, stronger authentication methods are required. To achieve this goal, it is expected that an authentication scheme that combines biometric techniques and existing authentication methods be studied further. In this paper, we propose a method to implement such biometric authentication based on gestures (dynamic characteristics) and body characteristics (static characteristics) and suggest ways to use them.

Keywords Biometrics · Gesture recognition · User authentication · Facial recognition

J. Oh
KEPCO, ICT Planning, Seoul, Republic of Korea
e-mail: jsun.oh@kepco.co.kr

J. Choi
Defense Information System Management Group, KIDA, Seoul, Republic of Korea
e-mail: jychoi@kida.re.kr

K. Moon
National Assembly, Seoul, Republic of Korea
e-mail: kyansik@assembly.go.kr

K. Lee (✉)
Graduate School of Information Security, Korea University, Seoul, Republic of Korea
e-mail: kevinlee@korea.ac.kr

© The Author(s), under exclusive license to Springer Nature Switzerland AG 2021
J. Kim and R. Lee (eds.), *Data Science and Digital Transformation in the Fourth Industrial Revolution*, Studies in Computational Intelligence 929,
https://doi.org/10.1007/978-3-030-64769-8_16

207

1 Introduction

As a conventional authentication method, recognizing a user by remembering a password or possessing an OTP generator (apparatus) has been widely used. Recently, biometrics (biometrics) has been introduced to strengthen the authentication means. Smartphone users can use PIN, pattern, fingerprint recognition, and face recognition (some devices) as a basic authentication method in a device security manner. Since personal information (photos, contact information, authentication information, and the like) are mainly stored in a smartphone device, an enhanced security authentication means should be provided. For personal information or data with restricted access, it is necessary to confirm that the user is an authorized user. In addition to password authentication, it is common to verify identity with OTP or biometric authentication as a secondary authentication.

In the case of PIN and OTP authentication methods, as a second-step authentication technology, the device must always be possessed (owner-based authentication), whereas in biometric authentication, the user himself becomes the authentication means and no other device is required. Also, while authentication information such as a password may be able to be hacked through several ways, biometric information is always changed each time at the time of authentication and has a characteristic that it is difficult to copy. In that respect, biometrics is recognized as an effective security authentication method. Recently the number of cyber-attacks on power institutes such as KHNP and KEPCO has reached to 1000 over five years. If even one attack succeeds, it can cause enormous damage such as paralyzing the power supply system. In this regard, ways to strengthen the access control are being studied, and the necessity of biometric authentication as a means of multi-factor authentication is highlighted [1].

This article introduces general authentication factors and biometric authentication, and overviews user authentication based on gestures and body characteristics as a dynamic biometric authentication method and attempts to describe the implementation scheme.

2 Background

In the past, personal and business relationships were formed based on human trust (user authentication era), but now, machines automatically process trust relationships instead (service authentication era). The first step in building these trusts is to handle right identification, which is called authentication. This section overviews authentication factors and technologies as a background theory and describes biometrics. We will also consider gesture recognition as a dynamic biometric method.

Table 1 The types of multi-factor authentications

Factors	Criteria	Characteristics	Examples
Knowledge factor	What you know	No device needed Easy to use Risk of forgetting	password, virtual keyboard image pattern
Ownership factor	What you have	Device required Risk of loss High security	IC card, OTP
Inherent factor	What you are	Recognition rate problem Authenticated by human presence Maintenance cost incurred High security	Biometrics

2.1 Authentication Factors

As a necessary condition for using an internet service, authentication is a technology for proving that the user is a legitimate user. It conceptually comes with three technical elements. The first element uses what the user knows as an authentication factor, the second uses the user's possession as an authentication factor, and the third uses the user's physical and behavioral characteristics. It can be seen separately according to the form used [2, 3]. That is, it can be classified into knowledge-based, possession-based, and presence(biological)-based authentication. And the authentication techniques can be implemented by combining three authentication factors with each other (Table 1).

2.2 Single-/Multi-Factor Authentication

Power control centers and national security research facilities often require higher level of authentication technology. This is because the ripple effect of damage is tremendous when a security breach occurs.

In the field such as financial services, they often require higher level of authentication technology. This is because security breaches are the most frequent and the ramifications of damage are high.

In these systems, user authentication may be implemented in various ways, but the most classic and frequently used method is ID and password authentication. In this method, the password is registered by the user first, and is encrypted in the service server using a hash algorithm and stored in the DB. When the service request is made, the user is confirmed to be a valid user by the password. Thus, authenticating only a single factor is referred to as single-factor authentication.

In contrast, multi-factor authentication uses a combination of two or more authentication factor technologies. For example, an additional popular OTP authentication device can be used to prove that the user is a legitimate user, who needs to go through a two-step process of passing OTP authentication again, after the successful password authentication [4].

2.3 Biometrics

2.3.1 Definition

Biometrics technology is to identify and authenticate an individual by extracting a user's physical and behavioral characteristics using a sensor device [5].

For this reason, the biometric authentication element must be unique to every person and must have a feature that can be used to quantify the extracted data using the sensor without any change.

For example, fingerprint and iris recognition correspond to this feature. Everyone has a fingerprint, but each is unique, making it easier to identify an individual and using a digital device to easily extract body features. Even before the use of digital sensors, it has been used to identify criminals using fingerprints in a classic way in criminal investigation.

2.3.2 Biometric Authentication Technology Classification

Biometric authentication technology can be distinguished by using physical and behavioral features [6, 7] (Fig. 1).

Physical Characteristics

In this method, authentication is performed using physical characteristics that are expressed by an individual body.

Fingerprints acquire a digital image using a dedicated sensor and recognize the user. They are classified into optical, capacitive, and ultrasonic fingerprints by the technology.

In the case of face recognition, it is a bio-recognition technique with the least resistance by human. And it is implemented by analyzing and recognizing each part having a characteristic face contour such as a nose, a mouth, eyebrows, and a chin in a mounted face.

In the case of iris recognition, it authenticates a user using the iris, which locates between the central black pupil and the white of the eye.

In the case of vein recognition, it uses the fact that is proven to show a different pattern for each person by using an after-glow by seeing a blood vessel with infrared rays and not being visible to the naked eye.

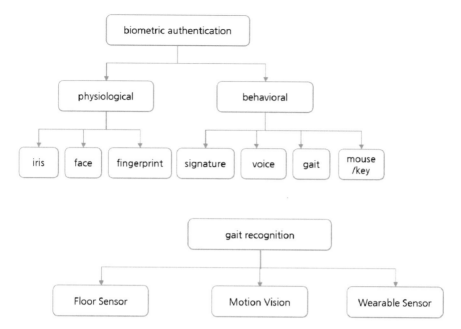

Fig. 1 Biometric authentication scheme can be classified by physiological and behavioral characteristics

Behavioral Characteristics

This method performs authentication using a specific pattern that is distinguished by the habit for everyone. There is voice recognition (speaker detection), walking (gait) recognition, and signature recognition, etc.

Speech recognition is a method of identifying a person based on phonetic characteristics, by extracting and analyzing features from a waveform of speech data extracted from a human voice.

Gait recognition is a method, in which a human footprint acts like a fingerprint, using a personal characteristic of a gait. This technique is being studied recently.

Besides, there is a method of extracting the handwriting pressure, the tilt of the pen, and the method of extracting the characteristics of the character shape.

Also, other research measures the brain waves of a person to identify the area activated when he is thinking, and this is used as an authentication factor.

2.4   Biometrics

2.4.1   Background

Gesture recognition-based user authentication is a technology for authenticating a user by automatically recognizing a person's face-gesture, hand-gesture, and other actions. At present, many research experiments have been conducted mainly on hand and face gestures [8].

2.4.2   User Authentication

Authentication process begins with the collection of information about identifiable features, which must be preceded by gesture recognition technology. It involves dynamically recognizing human gestures, which needs a sensor that can do it.

In the past, a method of attaching to the human body was studied, but nowadays, a depth camera (3D Depth Camera) is widely used, because it can be easily obtained, and it has advantages of versatility and convenience.

Gesture recognition using a depth camera analyzes data with a set of frames and tracking points that capture the motion. For example, in the case of Kinect, 30 frames are captured per second, and the points to be traced representing the movement of about 20 or more body-regions from the skeleton in the coordinate space (Fig.2).

The tracking point is where we are interested in biometric information, and we can use DTW (Dynamic Time Warping) [9] or HMM (Hidden Markov Model) as an algorithm to analyze this [10, 11].

These algorithms extract a unique pattern of behavior per each person, and we can determine the identity based on the similarity with the pattern at the time of authentication.

3   Implementation Method of Gesture Recognition-Based Authentication System

3.1   Overview

When it comes to implementing the biometric-based authentication system described above, it will be possible to enhance authentication and improve usability by combining multiple techniques.

The following figure shows a schematic diagram of a user authentication system that combines gesture recognition and other authentication means.

A representative example of the application of it is to enter a security zone. Generally, the entrance and exit of a national security facility requires the strongest level user authentication.

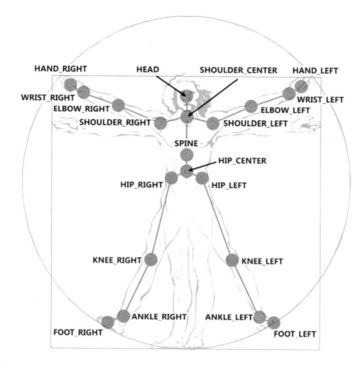

Fig. 2 Gesture recognition by using depth camera. (from Microsoft Kinect for Windows SDK)

First, a depth camera and screen terminal are installed at the entrance gate so that user input can be received. Additional authentication is performed based on information extracted from the user's wearable device, which can be added or disabled as needed.

After the first authentication, a method for improving user convenience by re-authenticating only the user's gesture, excluding terminal input, can be considered.

Until it is near the access gate, the wearable device collects and recognizes the user's biometric information (gait). The sensors (depth camera) are used for face recognition and voice recognition. Also, the touch of the screen(screen terminal), and the gesture of a mouse can be used.

Frequently, entrance and exit of nation security facilities use fingerprint recognition. As a means of strengthening authentication, it is possible to selectively combine these elements so that differential access control can be controlled according to the reinforced security policy.

3.2 Implementation of Dynamic, Static Biometrics

As described above, the target of biometric information recognition will be the touch screen of the terminal, usage pattern of the mouse, voice, facial-gesture, and hand-gesture.

Recognition of individual gestures and biometric information involves a common process of collecting information from individual sensors and authenticating through learning and inference processes.

If a personalized model is well learned through this process, unauthorized access to national security facilities cannot be bypassed due to the difficulty of imitation.

3.3 Implementation Method of Access Authentication System

3.3.1 Screen Touch Recognition

From Fig. 3, when touching the screen-terminal with a finger or palm, it is possible to consider a method to recognize the user's gesture by mapping the position touched on the screen to a two-dimensional coordinate plane and tracking the movement (Fig. 4).

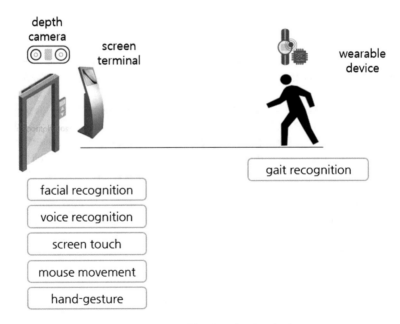

Fig. 3 Conceptual diagram of gesture recognition-based authentication system

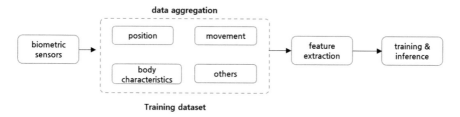

Fig. 4 A general mechanism that recognizes user behaviors, collects and learns data, and establishes an authentication model using biometric sensors

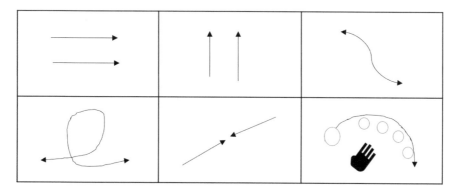

Fig. 5 Screen touch pattern examples

That method can be implemented, most easily, by the same way as the pattern-lock, used when unlocking the screen of the smart-phone. This is a method in which several screen touch patterns are defined, and a password is created by combining them. It is the most versatile and easy to implement. However, the difficulty may arises depending on the number of combinations (complexity) and the recognition rate of the touch screen.

Figure 5 shows some examples of the screen touch pattern. Since the degree of freedom is higher, as an extended authentication method, than the pattern-lock of the smart-phone, the benefit of enhancing security can be obtained.

3.3.2 Mouse Gesture Recognition

As with the screen touch method, a method for recognizing the unique movement pattern of the user using the mouse may be constructed in the same manner as the screen touch described above.

3.3.3 Hand Gesture Recognition

This method uses the information obtained by extracting the contours of the user's arms and hands from a short distance using the depth camera of Fig. 3, and to extract characteristic elements and learn a recognition model.

In implementation, two authentication model, using a static posture and using a sequence of continuous movements, can be considered.

In the former case, when a gesture meaning the same symbol is taken, the user is authenticated by distinguishing the static captured image.

In the latter case, a similarity between continuous motions is identified through a learning process that follows a predefined sequence of motions.

As a dynamic element of hand movement, continuous movements such as left and right movement, rotation, finger up and folding are patterned and recognized. But in actual use, learning process will be important. Such as displaying pre-defined gestures on the screen, and the user following them repeatedly until the accuracy reaches the minimum critical point.

3.3.4 Face Recognition

Face recognition has already been commercialized to a considerable area, with the rapid development of the field of image recognition.

We can consider two different methods, of statically extracting facial features or dynamically using facial gestures.

In the case of face recognition, since there is little resistance from user which requires minimal action from the user, it can be expected that the synergy effect of co-operation with others.

3.3.5 Gait Recognition

In the gait recognition, a camera or a wearable device can be used. In this paper, the wearable device will be described.

There are various biometric authentication factors using sensors of wearable devices, but this paper introduces an authentication method in connection with gait recognition [12]. When using a camera, it is necessary to install multiple cameras to shoot, which is difficult to be used for that recognition. Using a wearable device attached to the user solves such problems. The sensors used for the gait analysis can be various, and sensors such as an accelerometer and a gyroscope can be used. With the products released to the market, you can choose from a selection of built-in sensors that can be used for analysis.

The analysis will analyze the major joint movements, joint angles and forces and momentum. In addition, an electrocardiogram is measured, and the movement of a muscle is used, when gait is analyzed. In practice, it is possible to attach the device to the waist, pocket, or ankle for user authentication.

4   Application Scenario

Biometric authentication technology is rapidly growing with the embedding on smart phones and can be utilized in various fields with the expansion of IoT and wearable devices.

In the power industry, various service projects such as smart grid, smart distribution, remote monitoring of unmanned substation, and intelligent CCTV are being promoted. In order to perform these services securely and safely, biometric authentication technology can be introduced.

For example, when an authentication is required to obtain information in a transmission and distribution facility environment wearing insulated gloves, an alternative means can be utilized by recognizing an engineer's arm movement, which is convenient and safe to a person with limited behavior.

In addition, it can be used to provide a differentiated user experience. In the case of a facility engineer with a lot of outdoor work, gesture and voice authentication can provide more reliability and satisfaction.

5   Conclusion

In this paper, we described how to implement user authentication based on gestures and recognition of body characteristics as a biometric authentication.

Biometric authentication is unique to a user, difficult to duplicate, and characterized by non-identical authentication data each time, and is actively being used technology that can enhance security.

Static biometrics (such as fingerprints) are widely used today by recognizing the unique characteristics of the human-user but have the problem that it cannot be used if the fingerprint is duplicated or damaged.

Research on biometric authentication technology based on behavioral characteristics has been studied. In this paper, we have described how to recognize gestures for authentication.

In this paper, an authentication system for access control is considered as an example. The features are extracted from user-specific actions such as touching the screen, using a mouse, and hand movements, and user walking using wearable devices and authenticated for that user by pattern analysis.

In addition, it can be considered to construct a multi-level (multi-factor) authentication system by linking other technologies using face recognition and voice recognition. In actual use, it is possible to consider putting the difference on the level of combining the authentication according to authority, and to re-authenticate only by gait or some gestures to provide user convenience.

On the other hand, biometrics security standard or guideline will be needed for the procurement of it. It is expected that discussion will be made on the quantitative measure of security strength and indicators, and this will be left as a future research topic.

References

1. https://www.asiae.co.kr/article/2019092608122663584
2. Van Den Broek EL (2010) Beyond biometrics. Procedia Comput Sci 1(1):2511–2519
3. Apelbaum Y (2007) User Authentication principles, theory and practice. Fuji Technology Press
4. F. F. I. E. Council (2011) FFIEC releases supplemental guidance on internet banking authentication. DC: Federal Deposit Insurance Corp. (FDIC)
5. Mir A, Rubab S, Jhat Z (2011) Biometrics verification: a literature survey. Int J Computi ICT Res 5(2):67–80
6. Jain AK, Hong L, Pankanti S, Bolle R (1997) An identity-authentication system using fingerprints. Proc IEEE 85(9):1365–1388
7. Blackburn D, Miles C, Wing B (2009) Biometrics "foundation documents". Technical report, DTIC Document
8. Ducray B, Cobourne S, Mayes K, Markantonakis K (2015) Authentication based on a changeable biometric using gesture recognition with the kinect^{TM}. In: 2015 international conference on biometrics (ICB), pp 38–45. IEEE
9. Ten Holt G, Reinders M, Hendriks E (2007) Multi-dimensional dynamic time warping for gesture recognition. In: Thirteenth annual conference of the Advanced School for Computing and Imaging, vol 300
10. Huang X, Ariki Y, Jack MA (1990) Hidden Markov models for speech recognition. Edinburgh University Press
11. Rabiner LR (1989) A tutorial on hidden markov models and selected applications in speech recognition. Proc IEEE 77(2):257–286
12. Winter DA (1991) Biomechanics and motor control of human gait: normal, elderly and pathological. Waterloo Biomechanics Press: Waterloo, Ontario

AI TTS Smartphone App for Communication of Speech Impaired People

Hanyoung Lee and Deawoo Park

Abstract Due to COVID-19 in August 2020, the announcement by the Central Disaster and Safety Countermeasure Headquarters is being made regular every day. Sign language interpretation for the hearing impaired is broadcast on TV next to the speaker, but language communication for each individual for specific quarantine is insufficient. In this paper, AI construct motion recognition of sign language interpretation as Big Data, extract features of motion recognition data, and apply it to AI machine learning. Design an app to communicate with the hearing impaired on a smartphone owned by the hearing impaired. When a hearing impaired person writes TEXT or signs language on a smartphone, the smartphone recognizes it. Sign language motion is converted to text through motion recognition. The converted text is delivered to the general public as text to Speech through AI voice recognition. When an ordinary person speaks on a smartphone by voice, it is converted into text and displayed to the hearing impaired. This study will be used as a means of transmitting information to the hearing impaired in the Untact era. Develop a method for speech-impaired persons with disabilities to communicate with ordinary people who have not mastered sign language, using 4th industrial technology.

Keywords Artificial intelligence · Mobile · Big data · Motion recognition · Machine learning · TTS (Text to Speech) · STT (Speech to Text)

1 Introduction

The announcement by the government's Central Disaster and Safety Countermeasures Headquarters due to the Corona 19 (COVID-19) is being made on a daily basis through television and Internet media. With advanced fourth industrial revolution technology, wireless communication and internet are developed in areas of COVID-19 infected people to deliver information. A smartphone alone reveals the

H. Lee · D. Park (✉)
Department of Convergence Engineering, Hoseo Graduate School of Venture, Seoul, South Korea
e-mail: prof_pdw@naver.com

© The Author(s), under exclusive license to Springer Nature Switzerland AG 2021
J. Kim and R. Lee (eds.), *Data Science and Digital Transformation in the Fourth Industrial Revolution*, Studies in Computational Intelligence 929,
https://doi.org/10.1007/978-3-030-64769-8_17

movements of the COVID-19 confirmed patient and receives COVID-19 information through map search and AI voice search voice commands. At this time, disaster
measures are being delivered on TV through sign language for the deaf next to the
presenter of the Central Disaster and Safety Countermeasures Headquarters. The
obligation to deliver information to deaf people is important enough to be specified
by law.

However, communication is not easy for deaf people in real life. This is because
sign language interpreters must accompany them in order to communicate with non-
disabled people who have not learned sign language. As a result, communication
between deaf and non-disabled people is subject to great restrictions. In a rapidly
changing society, privacy is unprotected and only a limited number of people can
talk to them, so they can be placed in blind spots of information. In addition, AAC
(Augmentative and Alternative Communication), the existing supplementary alternative communication, was developed to meet the standards of people with develop-
mental disabilities who use picture cards of simple word combinations. Like Fig. 1,
there is an overwhelming majority over other types of disabilities, but the research
required for deaf people is insufficient. After meeting sign language interpreters and
hearing impaired people across the country, there is a need and demand for optimized
AAC that allows hearing impaired people to easily communicate with non-disabled
people [2]. This requires practical research to facilitate communication between the
deaf and the non-disabled.

In this paper, AI machine learning is applied to hearing-impaired people's smartphones to design APP for communication. When hearing-impaired people use TEXT
on their smartphones or sign language, their smartphones recognize their movements
and find the right words. In other words, sign language movements are converted to
TEXT through motion recognition. Translated TEXT is expressed as sound to non-
disabled people through TTS (Text to Speech). Conversely, when a non-disabled

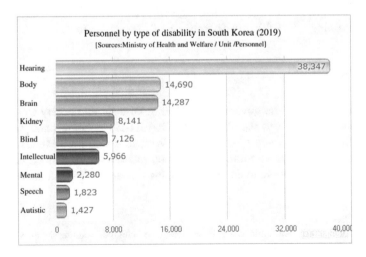

Fig. 1 Population by disability type in 2019 [1]

person speaks to a smartphone through voice, it converts it to TEXT (STT: Speech to Text) and shows it to deaf people.

This paper research will embody the methods of communicating with ordinary people who have not mastered sign language in the Untact era through AI, and will be used as a means of information communication and smooth communication.

2 Relevant Research

Sign Language Law: The sign language law aims to improve the language rights and quality of life of hearing-impaired and Korean sign language users by revealing that Korean sign language is the native language of hearing-impaired people with equal qualifications to the Korean language and laying the foundation for development [3].

Motion Recognition: Four basic elements of sign language (hand motion, hand shape, hand position, hand orientation) that communicate with speech-impaired people are identified. The motion recognition data stored in advance is compared using the similarity method, and the sign language is recognized [4]. Although lip motion, which is recognized by converting to 3D values to increase the accuracy of sign language, is best applied, lip motion cannot be developed into a smartphone application that is easy to use because it can be accompanied by special equipment such as two infrared cameras and three infrared LED (Light Emitting Diode)s. TTS (Text To Speech) & STT (Speech To Text): TTS refers to the conversion of letters, sentences, numbers, symbols, etc. into auditory information generally spoken by people, while STT refers to the conversion of voice into letters, numbers, symbols, etc. TTS technology has the advantage of being able to do both visual-dependent tasks. In addition, it is the easiest and easiest way to deliver information that can change from time to time, and it is a very efficient means of providing information for people with speech and hearing impairments [5].

AAC (Augmentative and Alternative Communication): AAC is a system with four components: symbols, tools, strategies, and techniques. It is divided into low-tech AAC [6] and high-tech AAC [7] f:by using this to communicate their opinions to others. Even the current state-of-the-art AAC uses symbols (picture cards) for the intellectually disabled, so there is a limit to having conversations for communication that the deaf want.

3 Motion Recognition of Deaf People Analysis and AI Design

3.1 Motion Recognition Analysis on Smartphones

With the launch of new products every few years, smartphones are convenient to use features that reflect the Fourth Industrial Revolution technology. You can search the Internet, play games, e-commerce, and financial transactions that can only be done on existing PCs, and you can call or search information using voice commands because AI voice recognition is possible.

In particular, IoT (Internet of Things) sensors that are detected in smartphones are becoming more intuitive to develop UI than before by using motion, direction, and heart rate check functions. Smartphones have become capable of various functions beyond the role of talking on the phone. Initially, it only detected simple movements through the camera, but now it can make big data for AI. Enter the result value into the repetitive motion of the person entering the camera to make it a meaningful action Fig. 2. In other words, data can be created by entering the sign language used by the deaf. Different people's hand gestures may make it difficult to extract accurate and meaningful words for each movement, but sign language is already used by people

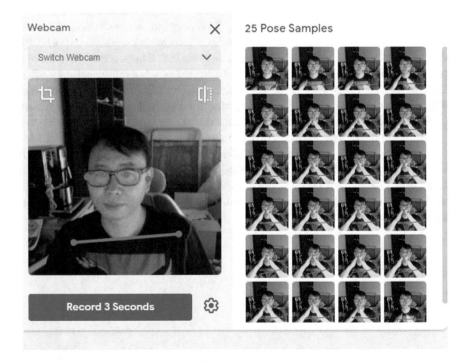

Fig. 2 Recognize 'home' through map learning using Google AI

Fig. 3 Guidance learning using AAC for the deaf

with speech and hearing impairments, so it forms the same pattern. Entering the sign language data of many people is a way to improve the accuracy of sign language interpretation, as input may vary depending on the size of a person's body type or movement Fig. 3.

You can use Google AI to learn sign language through instructional learning, and AAC allows you to enter your own sign language directly, so that you can save it in Big Data in the cloud and translate it accurately based on a number of learning samples. This is also available to those who wish to learn sign language, which could ultimately be an opportunity to break down the language barriers of the deaf and the non-disabled.

3.2 AI Machine Learning and TEXT Design

You can map and learn sign language movements on AI and build big data to understand the meaning of simple sign language movements and print them out in sentences. Through a combination of learned sign language words, a sentence can be completed by connecting the set words and TEXT the sentence. Figure 4 is

Fig. 4 Two different ways
of exemplar extraction

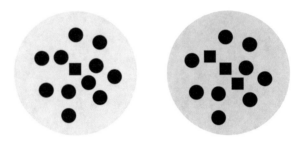

a comparison of the two representative extraction methods. The left side is how to select one single representative for each motion, and the right side is how to select multiple representatives.

Even if one representative is selected in the best way from a number of instructional data, in the end, from a different perspective, only one training data is used for each motion. To reduce the error of sign language, hydration by various deaf people is essential, and as more data builds up, more accurate words and sentences can be generated.

Complete the information collected by Sequence to Sequence, an Encoder-Decoder model using the RNN (Recurrent Neural Network).

In the case of proper nouns, consonants and vowels are recognized individually and the sign language as symbols is patterned as words, so it is only a list of meaningless words when data is first entered. In addition, due to the characteristics of sign language, investigations such as 'ya', 'ah' are omitted, so the consonants and vowels are combined in order of words and the corresponding surveys are attached to complete the string Fig. 5.

The encoder receives the data and compresses the information into a single vector. The vector at this point is called a context vector, and decoder uses this context vector to complete the sentence shown above. The disappointing point is that the longer the data words entered are listed, the less complete the combination is.

3.3 Designed with AI Machine Learning TEXT and TTS

Sentences produced by motion recognition will be printed in Text to Speech (TTS). Once a character identified by the image is entered, the process of processing the string is carried out. Locate the voice to match the subsequent string in syllables. It then leads to the process of finding syllable rhymes and creating speech fragments. Finally, the pieces entered into the image make a word and complete the sentence with the words, and each of these speech pieces is synthesized in order and then printed out.

The work has been advanced when delivering voice to TTS, but so far there are differences in each TTS program. The core of TTS is how accurately it recognizes text and reads it naturally without any awkwardness.

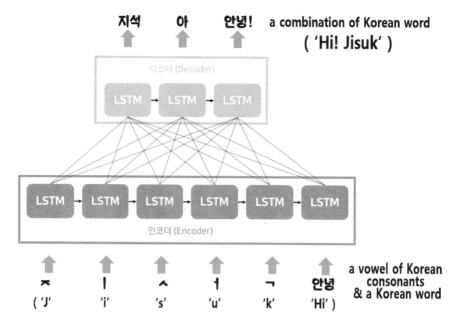

지석   아   안녕! a combination of Korean word
('Hi! Jisuk')

Fig. 5 Seq 2seq with full-connected layer in the middle

3.4 Designed with AI Machine Learning STT

The ability of non-disabled people to communicate with hearing-impaired people is limited to visual representation. One way to get out of that limitation is STT (Speech to Text).

Anyone can use STT easily with the GCP technology provided by Google. You can run gcloud and Google APIs in a shell window, or you can connect to Google Compute Engine (GCE) and run APIs.

4 AI App Design for AAC for the Deaf

4.1 Multi-AAC Development

AAC is a necessary skill for many types of disabled people who have difficulty communicating with their language. However, the high-tech AAC, which was developed as a symbol card that allows simple information to be exchanged with simple clicks, is less suitable for people with hearing impairments. Therefore, research and development of AAC devices suitable for hearing-impaired people who can use AI together with symbol cards for simple communication is carried out.

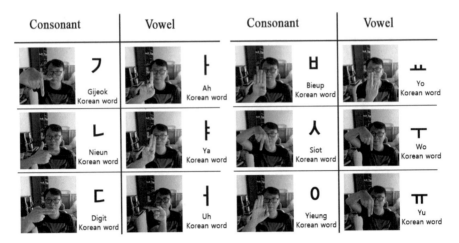

Fig. 6 Google AI sign language recognition via camera

4.2 Sign Language Recognition Analysis APP on Smartphones

The AAC, developed as a symbol card for deaf people to have a unique or deeper conversation, has a distinctive limitation. Google AI allows you to expand the scope of AAC, which is limited to recognizing and learning sign language Fig. 6.

4.3 AI Machine Learning String Learning Analysis APP

Converts to a string through a learned motion with an input image. Using RNN, complete the information collected by Seq 2seq with sentences and complete with TEXT. When the string is complete, it is delivered to non-disabled people by TTS Fig. 7.

4.4 AI Machine Learning Voice Learning Analysis & TEXT Display APP

In the process of recognizing voice and converting to TEXT, words are often omitted or distorted. This is due to the mechanical properties (micro sensitivity, ambient noise) that require continued development of hardware. STT has a difference in performance between a directional microphone and a microphone built into a smartphone. AI can improve accuracy by carrying out a number of studies through learning Fig. 8.

Fig. 7 Forward image data
learned by AI to TTS

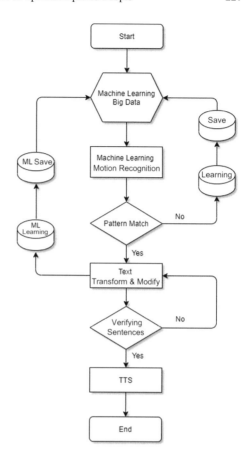

4.5 *Comparison of Information Transfer for the Deaf*

See Table 1.

5 Conclusion

To cope with social risk situations such as COVID-19 and disaster accidents, studies
on communication with deaf people are needed. Although existing technologies such
as TTS, STT, and motion recognition exist separately, they are not optimized for the
role of AAC for the deaf.

The research in this paper aims to develop auxiliary devices that can communicate with non-disabled and hearing-impaired people who cannot sign at all. In this
paper, the TTS STT technology is applied to the hearing impaired through the smartphone, combining IoT precedence and AI voice recognition learning technologies in

Fig. 8 Change the voice
learned by AI to TEXT

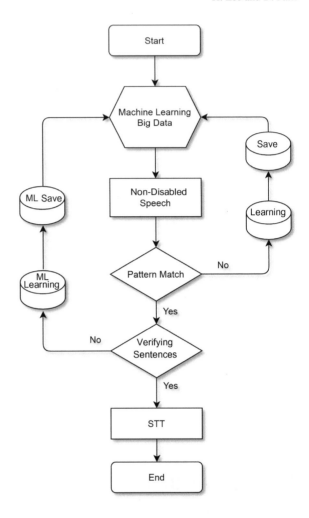

smartphones using the fourth industrial revolution technology. Through AI machine learning, a smartphone app was designed to learn the behavior recognition of sign language of the deaf and to accurately judge, and to design and suggest ways to use smartphones to communicate with non-disabled people. The study in this paper could protect the privacy of the deaf and have a positive effect on communication with the non-disabled due to the development of this device.

For future research, it is necessary to study ways to directly communicate normal people's language delivery from smartphones through hearing-impaired people's smartphones and directly using IoT hearing aids or user sensors.

Table 1 Comparison of AAC development for the deaf

	Existing	AI	Effect
AAC	By symbolic drawing alone limited in use	Communicable with sign language and symbols	Extended communication between the deaf and the non-disabled
Motion	Sign language recognition is inaccurate and serviceability is limited	Machine learning to recognize behavior and make more accurate sentences	Basic DB allows disabled people to teach themselves and provides learning opportunities for those who want to learn sign language.
TTS	In the context of pronouncing the exact words resulting in heterogeneity	Natural pronunciation	Natural conversation possible
STT	High missing words and error rates	Machine learning reduces word missing and typographical errors	Conveniently communicate with non-disabled people

References

1. datasom. http://www.datasom.co.kr/news/articleView.html?idxno=105467 (2020)
2. Hearing impaired; Gapyeong (Kim and 12 others) Sign language interpreter: (Lim and 12 others) (2020)
3. Korea Legal Information Center; http://www.law.go.kr/law/Korean Sign Language Method (2020)
4. Lee B-H, Lee K-S (2004) 2004 Recognition of finger language using image from PC camera. 102–104
5. Kim T, Kim B, Choi D, Lee Y (2012) Implementation of Korean TTS service on Android OS. 9–16 (2012)
6. Special Education Blog; https://m.blog.naver.com/bjs718/221797984500 (2020)
7. https://www.pacific.edu/about-pacific/newsroom/2018/october-2018/aac-camp-creates-space-for-communication.html (2018)
8. Heo D, Kim C, Kim S, Song B (2019) Comparative study on the speech recognition assistant device for participation in education of hearing impaired. 101–107
9. Kwon S, Choi Y, Lim S, Joung S (2016) A implementation of user exercise motion recognition system using smart-phone. 396–402
10. Yun T-J, Seo H-J, Kim D-H (2017) Text/voice recognition & translation application development using open-source. 426–427

Author Index

Printed in the United States
by Baker & Taylor Publisher Services